Springer Series in Computational Physics

Ralf Gruber Jacques Rappaz

Finite Element Methods in Linear Ideal Magnetohydrodynamics

With 103 Figures

Springer-Verlag
Berlin Heidelberg New York Tokyo

Dr. Ralf Gruber

Centre de Recherches en Physique des Plasmas, EURATOM Association,
Ecole Polytechnique Fédérale de Lausanne, 21, av. des Bains
CH-1007 Lausanne, Switzerland

Dr. Jacques Rappaz

Département de Mathématique, Ecole Polytechnique Fédérale de Lausanne
CH-1015 Lausanne, Switzerland

Editors

H. Cabannes

Mécanique Théorique
Université Pierre et Marie Curie
Tour 66-4, place Jussieu
F-75230 Paris Cedex 05, France

M. Holt

College of Engineering and
Mechanical Engineering
University of California
Berkeley, CA 94720, USA

H. B. Keller

Applied Mathematics 101-50
Firestone Laboratory
California Institute of Technology
Pasadena, CA 91125, USA

J. Killeen

Lawrence Livermore Laboratory
P.O. Box 808
Livermore, CA 94551, USA

S. A. Orszag

Department of Mechanical and
Aerospace Engineering
Princeton University
Princeton, NJ 08544, USA

V. V. Rusanov

Keldysh Institute of Applied Mathematics
4 Miusskaya pl.
SU-125047 Moscow, USSR

ISBN 978-3-642-86710-1 ISBN 978-3-642-86708-8 (eBook)
DOI 10.1007/978-3-642-86708-8

Library of Congress Cataloging in Publication Data. Gruber, Ralf, 1943–. Finite element methods in linear ideal magnetohydrodynamics. (Springer series in computational physics) Bibliography: p. Includes index. 1. Magnetohydrodynamics. 2. Finite element method. I. Rappaz, Jacques, 1947–. II. Title. III. Series. QC718.5.M36G78 1985 538'.6 85-4749

Softcover reprint of the hardcover 1st edition 1985

Monophoto typesetting, offset printing and bookbinding: Brühlsche Universitätsdruckerei, Giessen
2153/3130-543210

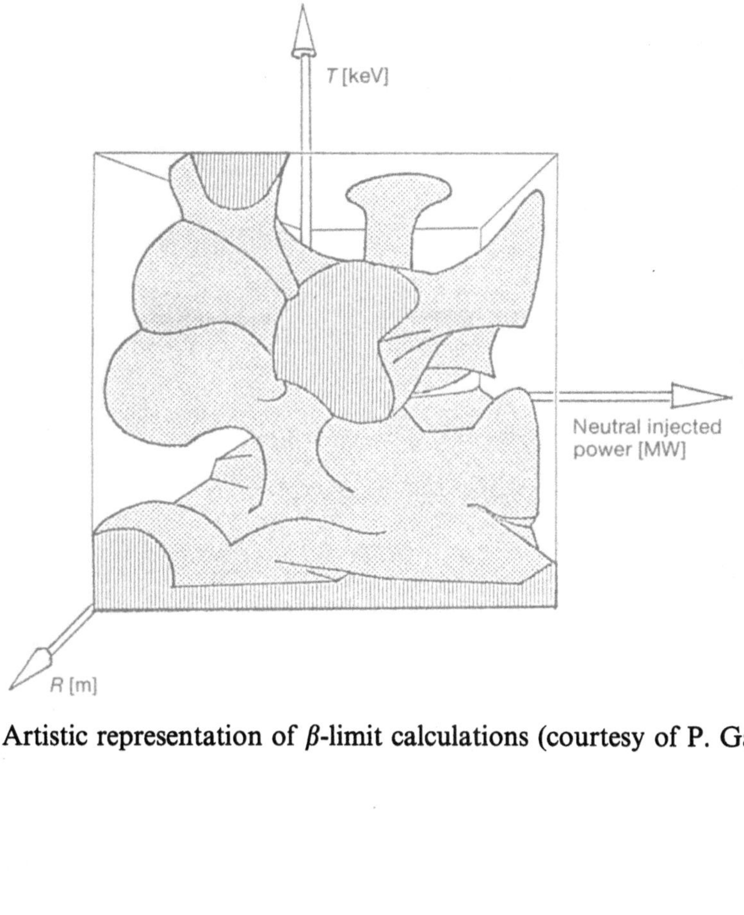

Artistic representation of β-limit calculations (courtesy of P. Gautier)

To the women of our lives

Ariane
Christine
Heidi
Hélène
Hulda
Kathrin
Sabine
Sylvie

Preface

For more than ten years we have been working with the ideal linear MHD equations used to study the stability of thermonuclear plasmas. Even though the equations are simple and the problem is mathematically well formulated, the numerical problems were much harder to solve than anticipated. Already in the one-dimensional cylindrical case, what we called "spectral pollution" appeared. We were able to eliminate it by our "ecological solution". This solution was applied to the two-dimensional axisymmetric toroidal geometry. Even though the spectrum was unpolluted the precision was not good enough. Too many mesh points were necessary to obtain the demanded precision. Our solution was what we called the "finite hybrid elements". These elements are efficient and cheap. They have also proved their power when applied to calculating equilibrium solutions and will certainly penetrate into other domains in physics and engineering.

During all these years, many colleagues have contributed to the construction, testing and using of our stability code ERATO. We would like to thank them here. Some of them gave partial contributions to the book. Among them we mention Dr. Kurt Appert, Marie-Christine Festeau-Barrioz, Roberto Iacono, Marie-Alix Secrétan, Sandro Semenzato, Dr. Jan Vaclavik, Laurent Villard and Peter Merkel who kindly agreed to write Chap. 6. Special thanks go to Hans Saurenmann who drew most of the figures, to Dr. Alan Turnbull for improving the English, to Barbara Kritz and Ellen Diethelm for the neat typewriting of the manuscript and to Pierre Gautier, now manager of a paper factory, for the artistic point of view of β optimization calculations. The work was led by Professor Francis Troyon who wrote a major part of Chap. 5. It was due to him that we were able to compile from many results a simple law which is now called the "Troyon limit".

These many results were partially produced on the CDC computers of the computing centre of our technical school. We express here our thanks to its staff who were even able to make many week-end arrangements possible for us. Also we are indebted to the computing centre of MPIPP of Garching. Without the use of their CRAY-1 the "Troyon limit" would never have been found.

It is a special pleasure to mention that a close collaboration between physicists and numerical analysts can lead to fruitful results. The work done by one of the authors (JR), by Prof. Jean Descloux and by Drs. Hugo Evequoz and

Yves Jaccard gave confidence in the applied methods. We encourage other research groups to get and keep in contact with numerical mathematicians. There is indeed a barrier of mutual comprehension to be overcome. If this is successfully done both parts can profit a lot. We hope that this book proves it.

Lausanne, February 1985

Ralf Gruber
Jacques Rappaz

Contents

Introduction

On the way towards a thermonuclear fusion reactor there are several technological and physical uncertainties to be understood and solved. One of the most fundamental problems is the appearance in the machines of many sorts of instabilities which can either enhance the energy outflow or even destroy the magnetic confinement of the fusion plasma. The knowledge of such instabilities is a prerequisite to a good understanding of the behaviour of actual experiments, and to the design of new devices. Most of the effort is devoted to the study of axisymmetric toroidal configurations such as tokamaks or spheromaks and to helically twisted toroidal devices such as stellarators.

One of the main objectives of the present fusion program is to show that nowadays tokamaks such as TFTR (Tokamak Fusion Test Reactor, USA) or JET (Joint European Torus, EURATOM) can reach temperatures, confinement times and β values ($=$ ratio between plasma pressure and the energy invested in the magnetic field) relevant to fusion reactors. The quantity β measures the efficiency of an electric power reactor based upon the tokamak principle. Different methods can be used to increase the value of β such as additional heating (neutral gas injection or radiofrequency heating), gas injection, decrease of the aspect ratio (minor to major radius), and increase of elongation or triangularity of the plasma cross section. Unfortunately, pushing the value of β too high leads to a desctruction of the plasma column which is called disruption, or to a degradation of the confinement time. It is thought today that there exist limits of β which depend on the geometry, the magnetic circuits and the properties of the fusion plasma.

The most violent global instabilities arising in timescales of microseconds are those described by the linear, ideal, magnetohydrodynamics MHD equations (*Kadomtsev* 1966). These are obtained when the linearized motion of a magnetically confined plasma around its equilibrium state is studied. In this model we neglect non-ideal effects such as the influence of finite resistivity, viscosity or kinetic effects. This model is well suited to reactor relevant plasmas having very high temperatures of the order of 10^8 K. For example, resistivity only alters the ideal results in timescales of the order of milliseconds, which is three orders of magnitude slower than the timescale of ideal modes. Thus, before including any non-ideal effect in stability considerations, we have to make sure that the ideal MHD model gives a stable equilibrium. Even though the model considered is simple and non-dissipative and the resulting eigenvalue problem symmetric, it is difficult to find general characteristics of the eigenvalue spectrum.

In the first stability investigations which have been performed analytically, one looked for stability criteria against localized modes in cylindrical and, later, in toroidal geometry. Many such stability criteria have been found. We cite here the most popular ones: *Suydam* (1958) derived a necessary criterion which tests the existence of unstable localized modes in cylindrical plasmas, and *Newcomb* (1960) gave necessary and sufficient conditions for such one-dimensional configurations. *Bineau* (1962) generalized the Newcomb criteria to toroidal configurations, and *Mercier* (1962), and *Greene* and *Johnson* (1962) found a necessary criterion for axisymmetric plasmas. More recently (*Connor* et al., 1978; *Dobrott* et al., 1977, and, for three dimensional configurations, *Correa-Restrepo* 1978), a stability criterion was derived for ballooning modes which are localized in radial and azimuthal directions. This criterion includes that found by *Mercier* (1962). The stability criterion for ballooning modes has been studied numerically by different groups. We only mention *Sykes* et al. (1983) who calculated optimal profiles for which the ballooning mode criterion was marginal everywhere in the plasma. This limit is now called "Sykes limit".

Quite a big effort has been made to derive stability criteria for global modes. As examples, we cite *Laval* et al. (1974) and *Dewar* et al. (1974) who studied the stability of the external kink modes in a straight column with elliptical cross section; and *Bussac* et al. (1975) who showed analytically the existence of a global internal unstable mode whenever there is enough shear and when the safety factor crosses 1 in the plasma region. They called this mode, "internal kink mode."

However, all these analytic calculations are not sufficient to understand the stability behaviour of global modes in general configurations starting from realistic equilibria. They give a very good insight into the mechanisms which drive the instabilities and thus help in the choice of the algorithms for the numerical calculation of general types of instabilities. Thus, during the seventies, many groups started to develop general purpose codes to study the stability of straight, axisymmetric, helical and three-dimensional configurations (*Bauer* et al. 1978; *Chodura and Schlüter* 1981; *Schlüter* and *Schwenn* 1981).

Many investigations on the stability of a straight cylindrical plasma column of circular cross section in which the equilibrium quantities depend only on one variable (the radius) have been made. Codes have been written which compute, for a given equilibrium, growthrates of instabilities and corresponding eigenfunctions. These codes use either a shooting method to solve the Hain-Lüst equation (1958), or a finite element method (*Takeda* et al. 1972; *Appert* et al. 1975). The finite element method has the advantage that it gives an overall view of the whole spectrum and thus permits a better qualitative understanding of the dependence of the eigenfrequencies and the eigenmodes on the parameters of the configuration.

The first successful codes to calculate the growth rates and the eigenfunctions in two-dimensional systems such as tokamaks or straight stellarators were the time-evolutionary codes (*Sykes* and *Wesson* 1974; *Bateman* et al. 1974).

In these calculations the linearized equations of motion are solved numerically by a finite difference scheme (*Richtmyer* and *Morton* 1967). An initial equilibrium state is perturbed randomly and the asymptotic time behaviour of this initial noise is determined. This technique works well in the high-β regime and gives reasonable accuracy for the most unstable kink mode. An excellent review of the results obtained with this method has been given by *Wesson* (1975). The time-evolutionary codes show resolution problems when treating tokamak-like plasmas with a low-β. For such plasmas the growth rate of the most unstable mode is very small, and the coupling of this mode with the marginally stable modes of the continuous spectrum makes it difficult to extract the unstable mode within a reasonable computing time. One also finds that localized modes cannot be obtained with the time evolutionary codes. This is probably due to the choice of the cylindrical coordinate system. Such a system does not fit the magnetic flux surfaces. As a consequence, a diffusion of the fluid elements through the flux surfaces takes place resulting in a stabilizing mode coupling. The main advantage of the time-evolutionary method is that the inclusion of dissipative effects is straightforward. It also constitutes the first step towards obtaining a fully non-linear and resistive code (*Sykes* and *Wesson* 1980; *Brackbill* 1976; *Hender* and *Robinson* 1981; *Finan III* and *Killeen* 1981; *Lynch* et al. 1981; *Bateman* 1978; *Edery* et al. 1981).

Another type of approach to the understanding of the ideal MHD stability problem are the semi-analytical codes. A surface current model is used in *D'Ippolito* et al. (1978) and a conforming mapping together with a high β tokamak expansion in *Goedbloed* (1981).

Finally, in a spectral code the time appears as an eigenvalue which is negative for a growing solution. The eigenfunction appearing in the variational form of the ideal linear MHD equations (*Bernstein* et al. 1958) is expanded in terms of a complete set of basis functions (*Zienkievicz* 1977; *Ciarlet* 1978; *Schwarz* 1980; *Strang* and *Fix* 1973) and substituted into the Lagrangian of the system. Different choices of basis functions can lead to different approaches. All these approaches use a global Fourier expansion in the ignorable coordinate which, for a tokamak, is the toroidal angle. In this direction the equilibrium solution is homogeneous and, as a consequence, the different wave numbers in the Fourier expansion decouple and can be considered one by one. In the two other directions, *Kerner* and *Tasso* (1975) (see also *Kerner* 1976) use a global Fourier expansion in the poloidal angle and a sophisticated, modified Bessel function expansion in the direction normal to the flux surfaces. The Princeton group use, in the original version of PEST (*Grimm* et al. 1976), also a global Fourier expansion in the poloidal angle and a finite element expansion in the normal direction. In the meantime they have written a new code, PEST 2 (*Manickam* et al. 1981), in which they test only the stability index defined by the sign of the eigenvalue. If the norm of the eigenvalue is not further given by the kinetic energy, it is possible to eliminate two of the three components of the displacement vector from the problem, and one remains with one partial differential equation similar to the Hain-Lüst equation (1958) for a one-dimensional geometry. In the ERATO code (*Gruber* et al. 1981a, b) described in detail in this book as well as in the GATO code (*Bernard* et al. 1981) the eigenfunction is expanded in two-dimensional non-conforming finite elements. All three spectral codes follow the same philosophy: They use sophisticated finite elements to obtain a spectral representation which is "unpolluted" (*Rappaz* 1976). Let us discuss this briefly:

In their first numerical paper, *Appert* et al. (1974a) saw that the spectrum of the ideal linear MHD equations in cylindrical geometry could not be reproduced well when using linear finite elements for all the three components of the displacement vector. In one class of eigenfunctions, instead of an infinitely degenerate eigenvalue a discrete spectrum which extended to infinity was obtained. This phenomenon was called "spectrum pollution". The reason for this pollution can be understood mathematically. Choosing a coordinate system and vector components such that two components are tangential to flux surfaces and one perpendicular to them, one can show that the radial dependence of the basis functions for the normal component has to be of one order higher than that for the tangential component. This condition is fulfilled when we choose basis functions in a space including the degenerate solution which fulfils $\nabla \cdot \xi = 0$, where ξ is the eigensolution. If we do so, the finite element approach is unpolluted. The unstable solutions which were strongly stabilized with the method used in *Appert* et al. (1974a) are then well represented and even the "eigenfunctions" corresponding to "eigenvalues" lying in the continuous spectrum obtain the expected δ-function like structures (*Appert* et al. 1975a).

Evequoz and *Jaccard* (1980) showed that the mathematical condition stated above holds for the 2D case as well. We have seen that it is most advantageous to choose conforming finite elements in a space including $\nabla \cdot \xi = 0$. This expansion scheme has been tested on a straight elliptical equilibrium (*Berger* et al. 1976). For a small ellipticity the external kink mode is well reproduced with a reasonable number of mesh cells. However, the weakly growing internal modes are very difficult to obtain, even in the most favorable circular case. This new difficulty can be traced to a poor representation of the $\boldsymbol{B} \cdot \nabla$ operator, which, in the one-dimensional case, is an algebraic quantity. This problem does not appear when using Fourier expansion in the poloidal angle direction (*Kerner* 1976; *Grimm* et al. 1976). In our stability codes ERATO and HERA (*Gruber* et al. 1981a, b) we solve this new problem by introducing non-conforming finite elements which we call "finite hybrid elements". The basic idea is to consider the vector components and their derivatives as distinct variables which are only identified at specific points in each mesh cell. The basis functions and the equilibrium quantities are chosen in such a way that each term in the Lagrangian is piecewise constant in each mesh cell. *Evequoz* (1980) has shown that this method is pollution free. The weakly growing modes are obtained already with few mesh cells in situations where, even with the highest resolution, it is impossible to find them with the conforming finite elements expansion. A quadratic convergence is observed from lower energy states (from below) in most of the tokamak cases examined so far. This destabilization of the lowest eigenvalue makes it possible to decouple it from the continuous spectrum, and enables us to obtain it with confidence. This finite hybrid element approach is very robust and guarantees that one does not miss the most unstable mode. The 2D stability codes ERATO (for toroidal geometry) and HERA (for helical symmetry) have reached a high level of maturity. Due to a poor representation of certain modes, parts of the code have been modified and rewritten many times. It is now possible to resolve eigenvalues corresponding to a timescale which is at the validity limit of the linear ideal MHD model. *Degtyarev* et al. (1984) have written a code which combines the ideas of PEST and ERATO. As in PEST they made a Fourier analysis in the toroidal and poloidal directions and expanded in finite hybrid elements in radial direction as done in ERATO.

For many years now, the most popular spectral codes PEST, ERATO and GATO (*Bernard* et al. 1981) have been applied to find optimal values of the plasma β in realistic geometries. In the first calculations performed by *Berger* (1977) and *Todd* et al. (1979), the optimization of β was made choosing the same functional forms for the pressure and current profiles, thus restricting too strongly the freedom of the parameters of the profiles. The obtained β values were of the order of 1%, too small for reactor relevant plasmas. These limits were rapidly overcome by experiments such as ISX–B (*Murakawi* et al. 1981). The next step towards high β values was to rely on a supraconducting shell, such that only internal modes had to be considered. As a consequence, β values of the order of 10% could be obtained by *Charlton* et al. (1979). The features of the internal modes have been discussed by *Kerner* et al. (1980), *Sykes* et al. (1983) and *Manickam* (1984). However, it was seen that the configurations were violently unstable against low and high toroidal wave number external modes. In the meantime more realistic equilibria have been considered. Also the restrictions in the choice of the functional forms of the profiles have been relaxed. As a consequence, much better agreement with experimental β limits can be found (*Charlton* et al. 1984; *Bernard* et al. 1983). Elaborate MHD stability studies have been performed by *Bernard* et al. (1980) and *Kerner* et al. (1981). In both studies an optimization procedure led to β values of 6% for JET and for a current of 10 MA. Since JET will never reach such currents this result can only have a theoretical impact. In the most recent calculations performed by *Troyon* et al. (1983) parameter studies have been performed for given total plasma currents. It was found that the optimal value of β scales with the total current and the inverse aspect ratio, and, for a given current, is independent of the geometry. A change in geometry only enables one to increase the total current. The scaling law corresponds well with measured optimal β values. (*Troyon* and *Gruber* 1984). These results are partially given in Sect. 5.7. The good agreement of these results with experiment leads one to suspect that the theoretical β limits are pertinent. This limit is now called "Troyon limit". *Degtyarev* et al. (1984) have found similar results. They first look for profiles which give stability with respect to ballooning modes. These profiles are then cut at the plasma surface to guarantee stability with respect to external kinks. The main difference to the other calculations is their higher β limit for a safety factor around 3 at the plasma surface. They also studied the influence of elongation and identation (bean shape tokamaks) on the optimal β values.

At present, there is a lively interest in the spheromak configuration (*Lüst* and *Schlüter* 1954; *Rosenbluth* and *Bussac* 1979) as an alternative confinement scheme to the tokamak concept. This configuration has the advantage that the toroidal field vanishes at the plasma surface and thus can be made compact, i.e. the aspect ratio of the torus can be 1. This advantage could predestinate this concept to become a fusion reactor. Stability calculations have been performed by a few groups (*Okabayashi* and *Todd* 1980; *Gautier* et al. 1981; *Finn* and *Reimann* 1982; *Jardin* 1982; *Pfersich* et al. 1983). The main outcome of these computations is that reducing the aspect ratio increases stability. It is believed that only the $n=1$ tilt mode remains unstable. The axisymmetric mode ($n=0$) is stable for an intermediate range in elongation. From these results, one sees that spheromaks could be an alternate solution to the tokamak concept.

Recently, MHD stability studies have been performed for helically symmetric equilibria without a net longitudinal current (straight stellerator). It was found (*Gruber* et al. 1981d; *Merkel* et al. 1983) that the stability limits of the external and internal modes coincide. The β limits found for a straight Heliotron (*Uo* et al. 1981) was 1% and for a straight Heliac (*Boozer* et al. 1982) β values up to $\beta=29\%$ were obtained. It must be remarked here that these two configurations are three-dimensional devices. The inclusion of the effect of the toroidal curvature, which was not considered in the above stability studies, could alter the obtained results. HERA was also used (*Charlton* and *Lee* 1983) to provide some information on the stability of the Advanced Toroidal Facility (ATF) project (*Lyon* et al. 1983).

The book is organized as follows: In Chap. 1 we introduce simple model equations which we solve with a variety of conforming and non-conforming finite element approaches. These model equations show the same numerical problems as those we met when solving the ideal linear MHD equations with the finite element method. Chapter 2 includes a brief presentation of the ideal MHD equations which define the static equilibrium state and the linear stability problem. In Chap. 3, we discuss the numerical problems which arise when solving the ideal linear MHD equations in cylindrical geometry numerically by the spectral code THALIA. One learns that "spectral pollution" is obtained when the basis functions are not taken in a certain function class. Conforming finite elements are first presented in Chap. 4 to solve the stability problem in a cylinder as a two-dimensional problem. The conforming elements have to be replaced by non-conforming finite hybrid elements to improve precision. In Chaps. 5 and 6 this finite hybrid element approach is applied to calculate β-limits in tokamaks and straight stellarators. The "spectral pollution" problem has been met in other different domains. Some of them are discussed in Chap. 7. Four appendices on the calculation of the ballooning mode criterion with finite elements, on an efficient eigenvalue solver making use of the specific matrix structures, on the organisation of ERATO and listings of the COMMONs used in ERATO and of ERATO 3 terminate the book.

1. Finite Element Methods for the Discretization of Differential Eigenvalue Problems

1.1 A Classical Model Problem

To familiarize those readers who mainly deal with finite differences rather than with finite elements, we present in this chapter the finite element approach to the variational form of the classical Sturm-Liouville eigenproblem on a bounded domain.

1.1.1 Exact Problem

Consider the equation

$$-\frac{d}{dx}\left[\alpha(x)\frac{du}{dx}(x)\right]+\beta(x)u(x)=\lambda\varrho(x)u(x) \quad \text{for} \quad x\in(0,1)^{1,2} \ , \tag{1.1}$$

subject to the *Dirichlet boundary condition* at $x=0$,

$$u(0)=0 \ , \tag{1.2}$$

and to the *Neumann boundary condition* at $x=1$,

$$\frac{du}{dx}(1)=0 \ . \tag{1.3}$$

The given functions α, β and ϱ in (1.1) are "sufficiently smooth" (for example in $C^\infty[0,1]^3$). We assume that $\alpha(x)>0$ and $\varrho(x)>0$ for all $x\in[0,1]$.

The problem we have to solve is to find real numbers λ and non-trivial functions u which satisfy (1.1–3). It is wellknown that a Sturm-Liouville problem such as this has an infinite sequence of real eigenvalues:

$$\lambda_1<\lambda_2<\ldots\lambda_j<\ldots\to\infty \tag{1.4}$$

---

1 (a, b) denotes the open interval of R with extremities a and b; $[a, b]$ is the closed interval with extremities a and b

2 $a\in A$ means; "a belongs to A"

3 $C^k[0,1]$ means the space of continuous functions with continuous derivatives up to and including order k on $[0,1]$. $C^\infty[0,1]$ means all derivatives are continuous on $[0,1]$

and an associated complete set of eigenfunctions $u_1, u_2, \ldots, u_j, \ldots$ satisfying the orthonormalization condition

$$\int_0^1 \varrho u_i u_j dx = \delta_{ij} . \tag{1.5}$$

Suppose that we have a real number λ and a function u satisfying (1.1–3). Multiplying (1.1) by any function $v \in C^1[0,1]$ with $v(0)=0$ and integrating it over the whole domain $[0,1]$, we obtain the formulation:

$$\int_0^1 \left[-v \frac{d}{dx}\left(\alpha \frac{du}{dx}\right) + \beta v u \right] dx = \lambda \int_0^1 \varrho v u \, dx . \tag{1.6}$$

The first term in (1.6) can be integrated by parts leading to

$$\int_0^1 \left(\alpha \frac{du}{dx}\frac{dv}{dx} + \beta u v \right) dx = \lambda \int_0^1 \varrho u v \, dx . \tag{1.7}$$

Here we have used

$$\alpha v \frac{du}{dx}\bigg|_0^1 = 0 , \tag{1.8}$$

since $v(0)=0$ and $du/dx(1)=0$.

We now ask from which space can u and v be chosen.

Consider $L^2(0,1)$, the set of all square integrable functions on $(0,1)$. If u and v belong to $L^2(0,1)$ we can define the scalar product

$$(u,v)_0 = \int_0^1 uv \, dx . \tag{1.9}$$

Let V be the set of all functions $w \in L^2(0,1)$ with $dw/dx \in L^2(0,1)$ and $w(0)=0$. Then V is a Hilbert space for the scalar product:[4]

$$(u,v)_1 = \left(\frac{du}{dx}, \frac{dv}{dx} \right)_0 + (u,v)_0 , \qquad u, v \in V . \tag{1.10}$$

We observe that the bilinear forms

$$a(u,v) = \int_0^1 \left(\alpha \frac{du}{dx}\frac{dv}{dx} + \beta u v \right) dx \quad \text{and}$$
$$b(u,v) = \int_0^1 \varrho u v \, dx \tag{1.11}$$

4 Space V is a Sobolev space and if $w \in V$ then w is a continuous function on $[0,1]$ (*Agmon* 1965)

are well defined for all $u, v \in V$. Equations (1.1–3) can now be formulated in the following way:

"Find real numbers λ and non-trivial $u \in V$ satisfying
$a(u, v) = \lambda b(u, v)$, for all $v \in V$." (1.12)

Note that the bilinear forms $a(u, v)$ and $b(u, v)$ in definitions (1.11) are symmetric in u and v. Problem (1.12) is called a *weak formulation* of (1.1–3).

Here we have to make an important remark: In the definition of the space V, we imposed the condition $u(0)=0$ (1.2) only. The *condition* $du/dx(1)=0$ (1.3) *is not imposed*. One can prove that if u satisfies (1.12) the condition (1.3) is *automatically fulfilled* (*Strang* and *Fix* 1973).

In other words:

The boundary condition $u(0)=0$ is *essential*.
The boundary condition $du/dx(1)=0$ is *natural*.

Let us introduce the *Rayleigh quotient*

$$R(u) = \frac{a(u, u)}{b(u, u)}, \quad u \in V . \tag{1.13}$$

Formulation (1.12) is equivalent to finding the *stationary or critical points of the functional* $R(u)$ on the space V. These are the points where the gradient of $R(u)$ vanishes (*Strang* and *Fix* 1973). For this reason, Problem (1.12) is also called a *variational formulation* of the eigenproblem defined by (1.1–3).

1.1.2 Approximate Problem

The variational formulation (1.12) of (1.1–3) is approximated by a finite element method. For this purpose, we first define a finite dimensional subspace V_h of V spanned by the basis functions $\phi_1, \phi_2, \ldots, \phi_N$, with N the dimension of V_h. A *Ritz-Galerkin method* for approximating the eigenelements of (1.12) consists of finding a real number λ_h and non-trivial functions $u_h \in V_h$, such that[5]

$$a(u_h, v_h) = \lambda_h b(u_h, v_h), \quad \text{for all} \quad v_h \in V_h . \tag{1.14}$$

If we expand

$$u_h(x) = \sum_{i=1}^{N} z_i \phi_i(x) , \tag{1.15}$$

where z_i are real numbers, (1.14) is equivalent to finding real numbers λ_h and vectors $z \neq 0$ with components $z_1, z_2, \ldots, z_N$, such that

$$\sum_{i=1}^{N} z_i a(\phi_i, \phi_j) = \lambda_h \sum_{i=1}^{N} z_i b(\phi_i, \phi_j), \quad j = 1, 2, \ldots, N . \tag{1.16}$$

5 It is usual to give an index h to quantities related to the approximate problem

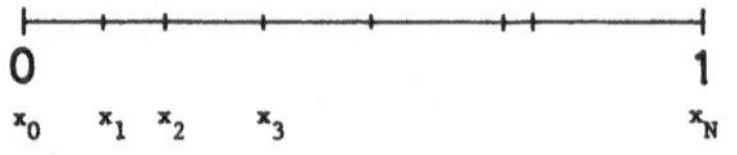

Fig. 1.1. Discretization of the domain $0 \leq x \leq 1$ in N intervals

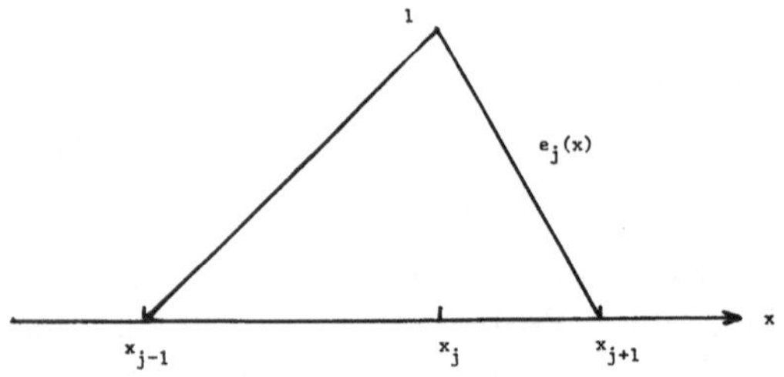

Fig. 1.2. Linear finite element $e_j(x)$ which is nonzero in the domain $x_{j-1} < x < x_{j+1}$, and zero elsewhere

Defining the matrices A and B such that the matrix elements are $a_{ji} = a(\phi_i, \phi_j)$ and $b_{ji} = b(\phi_i, \phi_j)$, (1.16) can be written as a matrix eigenvalue problem:

$$Az = \lambda_h Bz \ . \tag{1.17}$$

In (1.17), A and B are symmetric and B is positive definite [since $\varrho(x) > 0$]. This finite-dimensional eigenvalue problem can be solved by using a computer.

We have still to construct V_h. In practice V_h is generated by a set of basis vectors $\phi_1, \phi_2, \ldots, \phi_N$. It is advantageous to choose basis vectors with very small support. Such an approach is called a finite element method and leads to banded matrices A and B. The simplest choice for the ϕ_i is such that $u_h(x)$ in the expansion (1.15) becomes continuous and polynomial of degree $\leqq 1$ on each interval $[x_{i+1}, x_i]$, where $0 = x_0 < x_1 < x_2 < \ldots < x_{N-1} < x_N = 1$ are $(N+1)$ points chosen in the interval $[0, 1]$ (Fig. 1.1); we set $h = \max_i |x_{i+1} - x_i|$.[6] The function ϕ_i then becomes the hat function element which we shall call $e_i(x)$ and is defined by (Fig. 1.2):

for $i = 1, 2, \ldots, N-1$,

$$\begin{aligned} &e_i(x) = 0, \quad x \notin [x_{i-1}, x_{i+1}] \ , \\ &e_i(x) = (x - x_{i-1})/(x_i - x_{i-1}), \quad x \in [x_{i-1}, x_i] \ , \\ &e_i(x) = (x_{i+1} - x)/(x_{i+1} - x_i), \quad x \in [x_i, x_{i+1}] \ , \end{aligned} \tag{1.18}$$

and

$$\begin{aligned} &e_N(x) = 0, \quad x \notin [0, x_{N-1}] \ , \\ &e_N(x) = (x - x_{N-1})/(x_N - x_{N-1}), \quad x \in [x_{N-1}, 1] \ . \end{aligned} \tag{1.19}$$

6 There is an ambiguity in the meaning of h. It is, on the one hand, an index to specify the approximate problem, and on the other hand, it defines a mesh size. We will nevertheless go on with these common notations

$$A = 1/h \begin{pmatrix} 2 & -1 & & & & 0 \\ -1 & 2 & -1 & & & \\ & -1 & 2 & -1 & & \\ & & \ddots & \ddots & \ddots & \\ 0 & & & -1 & 2 & -1 \\ & & & & -1 & 1 \end{pmatrix} \qquad B = h/6 \begin{pmatrix} 4 & 1 & & & & 0 \\ 1 & 4 & 1 & & & \\ & 1 & 4 & 1 & & \\ & & \ddots & \ddots & \ddots & \\ 0 & & & 1 & 4 & 1 \\ & & & & 1 & 2 \end{pmatrix}$$

Fig. 1.3. Matrices A and B for the classical model problem

In the one-dimensional problem (1.1–3), such a choice of basic functions leads to tridiagonal matrices A and B. If we choose $\alpha(x)=\varrho(x)=1$ and $\beta(x)=0$ for all $x\in[0, 1]$ and for an equidistant mesh ($h=|x_{i+1}-x_i|$ for all i), we obtain the matrices A and B given in Fig. 1.3. Note that the integration (1.11) for the matrix elements can be performed analytically. However, in more general cases, this integration has to be performed numerically.

If we discretize (1.1–3) by finite differences we would obtain the same matrix A. However, matrix B would then be a diagonal matrix and the matrix eigenvalue problem, (1.17), would be easier to solve. This advantage of finite differences over finite elements – we shall see that finite elements have many advantages over finite differences – is often used in the field of structural mechanics by doing what is referred to as "mass-lumping" which consists of replacing B in Fig. 1.3 by $B=hD$, where D is a diagonal matrix. In fact, D can be obtained by applying a trapezoidal rule for the integration in (1.11).

1.1.3 Questions on Numerical Stability

When replacing the exact problem (1.12) by the approximate one, (1.14), a few questions arise:

I) If λ is an eigenvalue of Problem (1.12), do eigenvalues λ_h of (1.14) exist such that $\lim_{h\to 0}\lambda_h=\lambda$?

II) Are the eigenvalues $\lambda=\lim_{h\to 0}\lambda_h$ solutions of Problem (1.12)?

III) Do the eigenfunctions u_h of (1.14) tend towards the function u of (1.12)?

IV) Is it possible to obtain error estimates between the eigenelements of (1.12) and those of (1.14)?

For the classical model problem, the answer to all four questions is positive (*Strang* and *Fix* 1973). To illustrate this, we solve the eigenvalue problem (1.17), as a function of h. The functions $\alpha(x)$, $\beta(x)$ and $\varrho(x)$ are chosen to be 1. The result is shown in Fig. 1.4. The straight lines towards the analytic solutions ($h=0$),

$$\lambda_k=1+\left(\frac{\pi}{2}+k\pi\right)^2 \qquad (1.20)$$

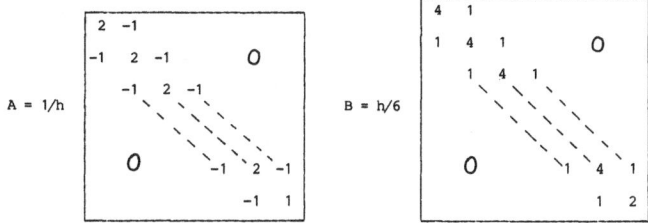

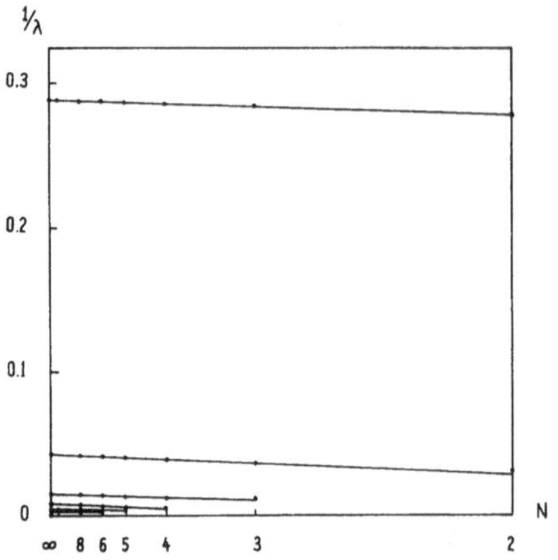

Fig. 1.4. Numerical spectrum of the classical model problem as a function of N. The eigenvalue of each mode converges quadratically to the analytic solution

and

$$u_k = \sin\left(\frac{\pi}{2} + k\pi\right)x, \qquad k = 0, 1, 2, \dots \tag{1.21}$$

reveal a quadratic convergence behavior.

In the following example, we will see that for non-standard (non-classical) problems, the answer to question (II) can be negative. In that case, we can say that the approximated problem (1.14) introduces a "pollution effect".

1.2 A Non-Standard Model Problem

1.2.1 Exact Problem

Consider the system of ordinary differential equations

$$\begin{aligned} -\frac{d}{dx}\left(\alpha\frac{du_1}{dx}\right) - \frac{d}{dx}(\beta u_2) + \gamma u_1 &= \lambda\varrho u_1 \ , \\ \beta\frac{du_1}{dx} + \delta u_2 &= \lambda\varrho u_2 \ , \end{aligned} \tag{1.22}$$

on the interval (0, 1), subject to the boundary conditions

$$\begin{aligned} u_1(0) &= 0 \ , \\ \alpha(1)\frac{du_1}{dx}(1) + \beta(1)u_2(1) &= 0 \ . \end{aligned} \tag{1.23}$$

The functions α, β, γ, δ and ϱ are of class $C^\infty[0, 1]$ with $\alpha(x) > 0$ and $\varrho(x) > 0$ for all $x \in [0, 1]$. The problem is to find real numbers λ and non-trivial functions

$\boldsymbol{u}=(u_1, u_2)$ satisfying (1.22 and 23). Choosing first

$$\begin{aligned} &\alpha(x)=\beta(x)=\delta(x)=\varrho(x)=1 \ , \\ &\gamma(x)=0, \quad \text{for all} \quad x\in[0,1] \ , \end{aligned} \tag{1.24}$$

equations (1.22 and 23) become

$$\begin{aligned} &-\frac{d^2u_1}{dx^2}-\frac{du_2}{dx}=\lambda u_1 \ , \\ &\frac{du_1}{dx}+u_2=\lambda u_2 \ , \end{aligned} \tag{1.25}$$

and

$$\begin{aligned} &u_1(0)=0 \ , \\ &\frac{du_1}{dx}(1)+u_2(1)=0 \ . \end{aligned} \tag{1.26}$$

One can see that $\lambda=0$ and $\boldsymbol{u}=(w, -dw/dx)$ is the solution of (1.25 and 26) where w can be any "sufficiently regular" function with $w(0)=0$. In this case the eigenvalue $\lambda=0$ is *infinitely degenerate* since the eigenspace corresponding to $\lambda=0$ is of infinite dimension. Eliminating u_2 in (1.25) leads to

$$-\frac{d^2u_1}{dx}+u_1=\lambda u_1 \ , \tag{1.27}$$

which is identical to Problem (1.1) with $\alpha=\beta=\varrho=1$. The eigenvalues are given by (1.20) and the eigenvectors by (1.21).

Let us consider a second particular case for which we choose

$$\begin{aligned} &\alpha(x)=1+x, \quad \beta(x)=\delta(x)=\varrho(x)=1 \ , \\ &\gamma(x)=0, \quad \text{for all} \quad x\in[0,1] \ . \end{aligned} \tag{1.28}$$

Equations (1.22 and 23) then become

$$\begin{aligned} &-\frac{d}{dx}\left[(1+x)\frac{du_1}{dx}\right]-\frac{du_2}{dx}=\lambda u_1 \ , \\ &\frac{du_1}{dx}+u_2=\lambda u_2 \ , \end{aligned} \tag{1.29}$$

with

$$\begin{aligned} &u_1(0)=0 \ , \\ &2\frac{du_1}{dx}(1)+u_2(1)=0 \ . \end{aligned} \tag{1.30}$$

In this case the infinitely degenerate eigenvalue ($\lambda=0$) of the first problem, (1.25 and 26), opens to a *continuous spectrum*[7] in the interval $[0, 1/2]$. This can be seen when we eliminate u_2 in (1.29). In the remaining equation for u_1,

$$-\left(\frac{\lambda}{\lambda-1}+x\right)\frac{d^2u_1}{dx}-\frac{du_1}{dx}=\lambda u_1 \ . \tag{1.31}$$

The coefficient

$$g(x)=\frac{\lambda}{\lambda-1}+x \tag{1.32}$$

vanishes at some point $x\in[0,1]$ for $\lambda\in[0,1/2]$, and consequently the differential operator of (1.31) does not admit a continuous inverse in $L^2(0,1)$.

The non-standard model problem (1.22, 23) may contain an infinitely degenerate eigenvalue, a continuous spectrum or an accumulation point at some finite value of λ. This is not the case in the standard Sturm-Liouville problem, (1.1–3). In this sense, Problems (1.1–3) and (1.22, 23) are quite different ones. We will see later that the eigensystems given by the ideal linear magnetohydrodynamic equations describing the stability behavior of a magnetically confined ionized gas are of the same type as those in Problem (1.22, 23).

To obtain the variational formulation of (1.22, 23), we multiply the first equation by a smooth function v_1, with $v_1(0)=0$, and the second equation in (1.22) by v_2, add them up and integrate over $[0,1]$. After integration by parts of the first two series, one obtains

$$\int_0^1\left(\alpha\frac{du_1}{dx}\frac{dv_1}{dx}+\beta u_2\frac{dv_1}{dx}+\gamma u_1v_1+\beta\frac{du_1}{dx}v_2+\delta u_2v_2\right)dx$$
$$=\lambda\int_0^1\varrho(u_1v_1+u_2v_2)dx \ . \tag{1.33}$$

In order to obtain the same formulation (1.12) as for the standard model problem, we again define

$$a(\boldsymbol{u},\boldsymbol{v})=\int_0^1\left(\alpha\frac{du_1}{dx}\frac{dv_1}{dx}+\beta u_2\frac{dv_1}{dx}+\gamma u_1v_1\right.$$
$$\left.+\beta\frac{du_1}{dx}v_2+\delta u_2v_2\right)dx \ , \tag{1.34}$$

and

$$b(\boldsymbol{u},\boldsymbol{v})=\int_0^1\varrho(u_1v_1+u_2v_2)dx \ .$$

7 The exact definition of a continuous spectrum will be given in Sect. 1.3

Also we introduce a new space V which is the set of all functions $\boldsymbol{u}=(u_1, u_2)$ such that $u_1 \in L^2(0,1)$, $du_1/dx \in L^2(0,1)$, $u_1(0)=0$ and $u_2 \in L^2(0,1)$. The variational formulation of (1.22, 23) then is:

"Find real numbers λ and non-trivial functions $\boldsymbol{u}=(u_1, u_2) \in V$ such that $a(\boldsymbol{u}, \boldsymbol{v})=\lambda b(\boldsymbol{u}, \boldsymbol{v})$, for all $\boldsymbol{v}=(v_1, v_2) \in V$." (1.35)

As in formulation (1.12) the bilinear forms $a(\boldsymbol{u}, \boldsymbol{v})$ and $b(\boldsymbol{u}, \boldsymbol{v})$ are symmetric in $\boldsymbol{u}$ and $\boldsymbol{v}$. Moreover, we remark that there are no derivatives on u_2 and v_2. The boundary condition $u_1=(0)$ is an *essential boundary condition* included in the definition of space V and

$\alpha(1)du_1/dx(1)+\beta(1)u_2(1)=0$ is a *natural boundary condition.*

1.2.2 Conforming "Polluting" Approximations

As in Sect. 1.1.2, we build a family of finite-dimensional subspaces V_h of V and solve the problem,

"Find real numbers λ_h and non-trivial functions $\boldsymbol{u}_h \in V_h$ such that $a(\boldsymbol{u}_h, \boldsymbol{v}_h)=\lambda_h b(\boldsymbol{u}_h, \boldsymbol{v}_h)$, for all $\boldsymbol{v}_h \in V_h$." (1.36)

We discretize the domain $[0,1]$ into N equidistant intervals ($h=1/N$) and denote by $x_j=jh$, $j=0,1,\ldots,N$ the positions of the mesh points (Fig. 1.1). For the general basis functions ϕ_j we choose again the particular hat basis functions $e_j(x)$ as they are given by (1.18, 19) and represented in Fig. 1.2. In addition we have to give the hat basis function $e_0(x)$ attributed to the first point $x_0=0$:

$$\begin{aligned} e_0(x)&=0 && \text{if} \quad x\in[x_1,1] \;, \\ e_0(x)&=\frac{x_1-x}{x_1-x_0} && \text{if} \quad x\in[0,x_1] \;. \end{aligned} \tag{1.37}$$

We expand $\boldsymbol{u}_h \in V_h$:

$$\boldsymbol{u}_h(x)=(u_{1h}(x), u_{2h}(x))=\left(\sum_{i=1}^{N} p_i e_i(x), \sum_{i=0}^{N} q_i e_i(x)\right) \;, \tag{1.38}$$

where p_i and q_i are real numbers describing the function values of $u_{1h}(x_i)$, and $u_{2h}(x_i)$, respectively. Again we find an eigenvalue problem of the type (1.17) with the matrix elements

$$\begin{aligned} a_{ji}=a(\phi_i, \phi_j) \quad \text{with} \quad &\phi_i=(e_i, 0), && 1 \leqq i \leqq N \;, \\ &\phi_{i+N+1}=(0, e_i), && 0 \leqq i \leqq N \;. \end{aligned}$$

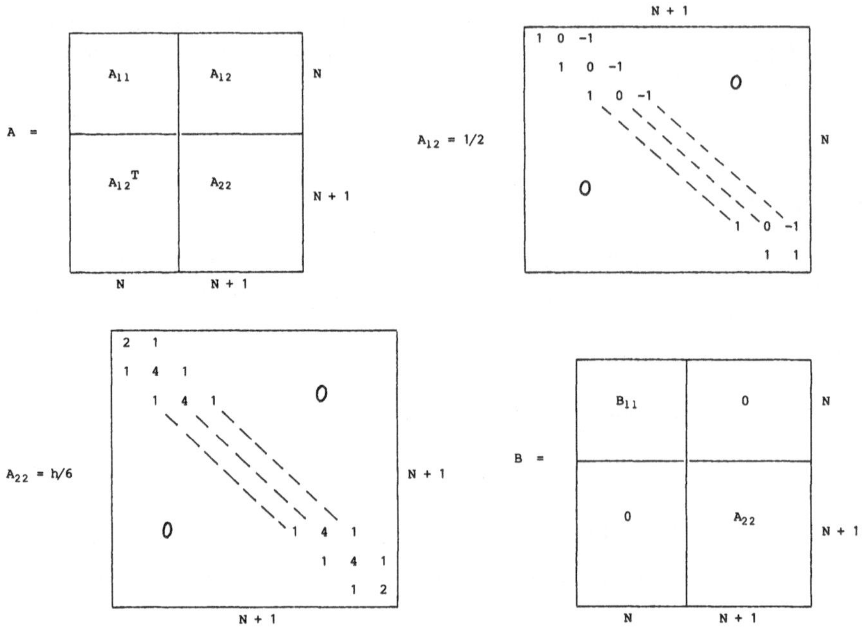

Fig. 1.5. Matrices A and B for the non-standard model problem with a degenerate eigenvalue $\lambda=0$. The submatrices A_{11} and B_{11} are the same as A and B in Fig. 1.3, respectively

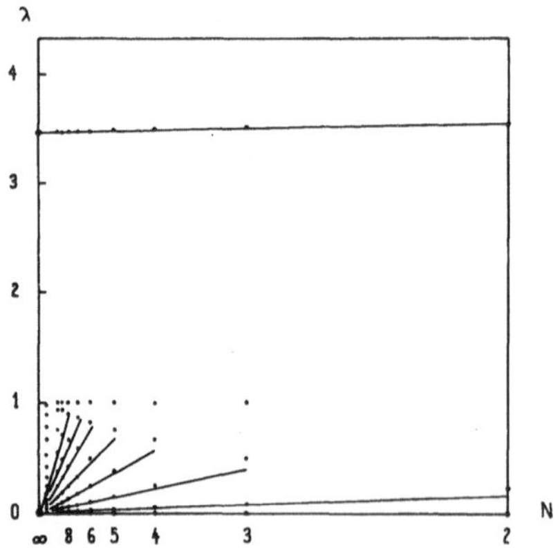

Fig. 1.6. Numerical spectrum of the non-standard model problem as a function of N. Linear elements for u_1 and u_2 are used. The discrete modes are well reproduced. Instead of the degenerate eigenvalue $\lambda=0$, one obtains a spectrum in the domain $0<\lambda<1$. The eigenvalue of each mode with a fixed radial wave number converges quadratically towards $\lambda=1$ as N is increased. This phenomenon is called "spectral pollution"

This hat finite element approach for both components u_1 and u_2 is now applied to our two non-standard model problems. The matrices A and B for the first Problem, (1.25, 26), are shown in Fig. 1.5. The eigenvalues as a function of the mesh size h are represented in Fig. 1.6. Straight lines give quadratic convergence. At the $h=0$ axis we give the analytic solution, which consists of the infinitely degenerate eigenvalue $\lambda=0$ and the Sturm-Liouville solution

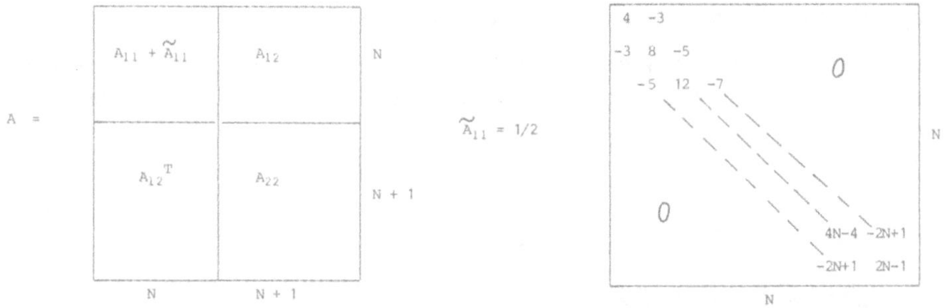

Fig. 1.7. Matrix A for the non-standard model problem with a continuum using linear finite elements for the discretization of u_1 and u_2. The submatrices A_{11}, A_{12} and A_{22} are the same as in Fig. 1.5

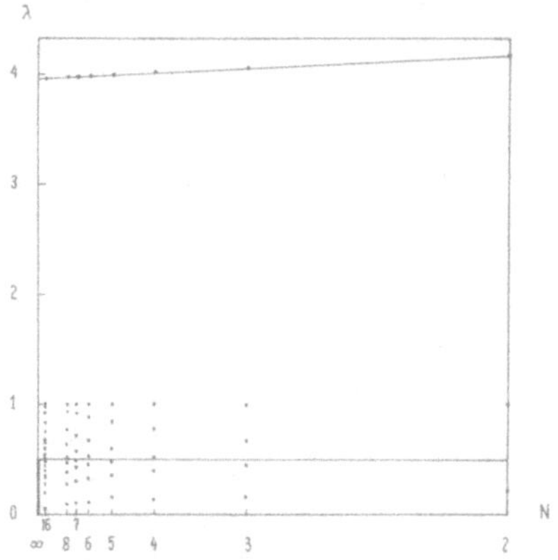

Fig. 1.8. Spectral pollution in the presence of a continuous spectrum covering the domain $0 \leq \lambda \leq 0.5$. Numerically the "continuum" extends from $\lambda = 0$ to $\lambda = 1$

given in (1.20). We see that the Sturm-Liouville solution is well represented. The degenerate solution $\lambda = 0$ is wrongly represented. If N grows, the highly oscillating eigenfunctions corresponding to the eigenvalue $\lambda = 0$ fill in more and more the domain $\lambda \in [0, 1]$. The question (II) in Sect. 1.1.3 cannot further be answered by yes. This phenomenon of spreading an infinitely degenerate eigenvalue is proposed to be called "spectral pollution".

A very similar effect is seen when discretizing (1.29, 30) by a set of hat finite elements. The corresponding matrix A is shown in Fig. 1.7. In Fig. 1.8 we again represent λ as a function of the mesh size, h, such that a quadratic convergence can be detected from straight lines. At the $h = 0$ axis we represent the exact solution, which consists of a continuous spectrum in the domain $\lambda \in [0, 1/2]$ and a Sturmian part. As in the first non-standard case we observe a "spectral pollution" and question (II) in Sect. 1.1.3 is again answered by no.

However, there exist finite element subspaces V_h which do lead to a positive answer.

1.2.3 "Non-Polluting" Conforming Approximation

Let $N>0$ be an integer. We set $h=1/N$ and $x_i=ih$; $i=0, 1, \ldots, N$ (Fig. 1.1). The set of *piecewise constant functions* $c_{i-1/2}$ in each interval $[x_{i-1}, x_i]$ is generated by:

$$
\begin{aligned}
c_{i-1/2}(x)&=0 \quad \text{if} \quad x \notin [x_{i-1}, x_i] \ , \\
c_{i-1/2}(x)&=1 \quad \text{if} \quad x \in (x_{i-1}, x_i) \ .
\end{aligned}
\tag{1.39}
$$

We expand $\boldsymbol{u}_h \in V_h$ (Fig. 1.9):

$$
\boldsymbol{u}_h(x)=(u_{1h}(x), u_{2h}(x))=\left(\sum_{i=1}^{N} p_i e_i(x), \sum_{i=1}^{N} q_{i-1/2} c_{i-1/2}(x)\right) \ , \tag{1.40}
$$

where p_i and $q_{i-1/2}$ are real numbers describing the function values of $u_{1h}(x_i)$ and $u_{2h}(x_{i-1/2})$, respectively; the position $x_{i-1/2}=(x_{i-1}+x_i)/2$. The basis functions e_i are defined in (1.18, 19). The space V_h is a finite-dimensional subspace of V.

This *hat finite element approach for the first component* u_1 *and the piecewise constant element approach for the second component* u_2 is applied to our two non-standard model problems. The matrices A and B for the first one, (1.25, 26), are given in Fig. 1.10. The eigenvalues as a function of the mesh size, h, are plotted in Fig. 1.11. The spectrum is well represented. With N intervals we obtain N degenerate modes corresponding to $\lambda=0$. No pollution effect is seen. Question (II) can be answered positively.

What happens with a continuous spectrum? We discretize the second non-standard problem, (1.29, 30), using expansion (1.40). The matrix A is shown in Fig. 1.12. The eigenvalues as a function of the mesh size h are plotted in Fig. 1.13. One observes that the domain $\lambda \in (0, 1/2)$ is filled in with increasing N. For these "continuous modes" it is not further possible to talk about quadratic convergence. What happens is that the density of eigenvalues increases with increasing N. Each eigenvalue is related to the position of a mesh cell and the eigenfunction is "close" to a δ function at this point. Changing N changes the

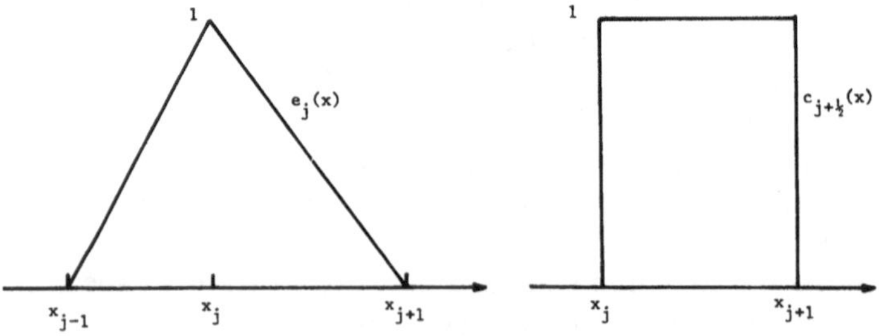

Fig. 1.9. Linear finite element $e_j(x)$ for the discretization of u_1 and piecewise constant element $c_{j+1/2}$ for u_2

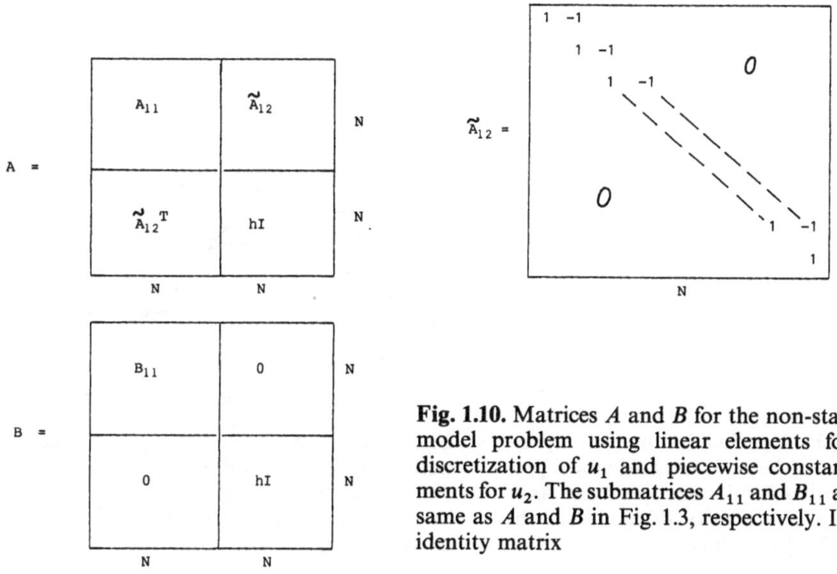

Fig. 1.10. Matrices A and B for the non-standard model problem using linear elements for the discretization of u_1 and piecewise constant elements for u_2. The submatrices A_{11} and B_{11} are the same as A and B in Fig. 1.3, respectively. I is the identity matrix

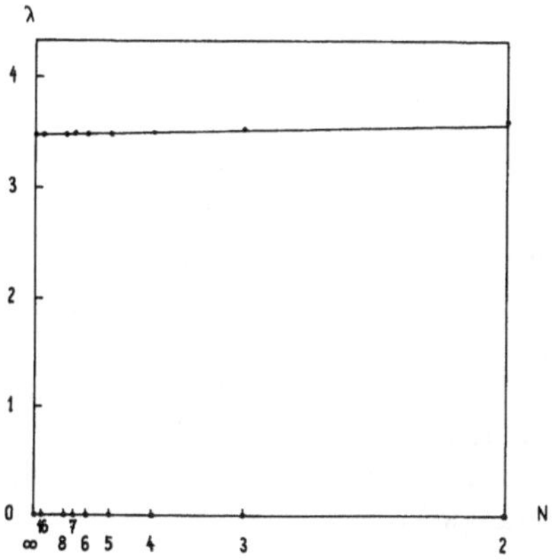

Fig. 1.11. Numerically calculated spectrum of the non-standard model using different elements for u_1 and u_2. The eigenvalue $\lambda = 0$ is obtained with the expected $N-1$ fold degeneracy

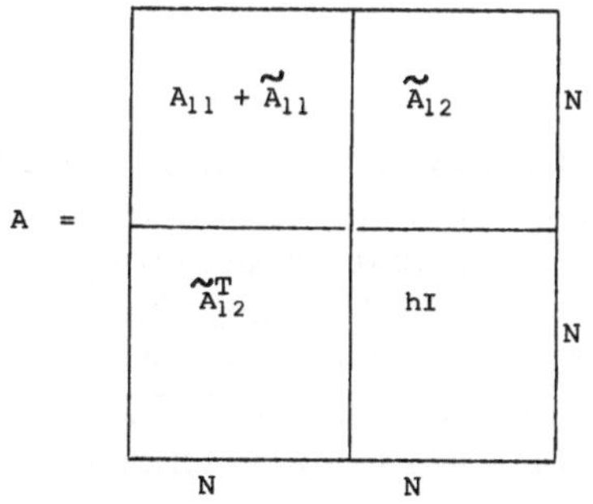

Fig. 1.12. Matrix A for the non-standard model problem with a continuous spectrum using linear finite elements for the discretization of u_1 and piecewise constant elements for u_2. Submatrices A_{11}, $\tilde{A}_{11}$ and $\tilde{A}_{12}$ are those given in Figs. 1.3, 1.7 and 1.10, respectively

position of most mesh cells and, as a consequence, changes the eigenvalue distribution of the continuous spectrum.

This "unpolluted" approach will be justified mathematically in Sect. 1.3. However, we can give here a few heuristic arguments on why such an approach is well suited for the problem (1.25, 26). Consider the second equation in (1.25) in

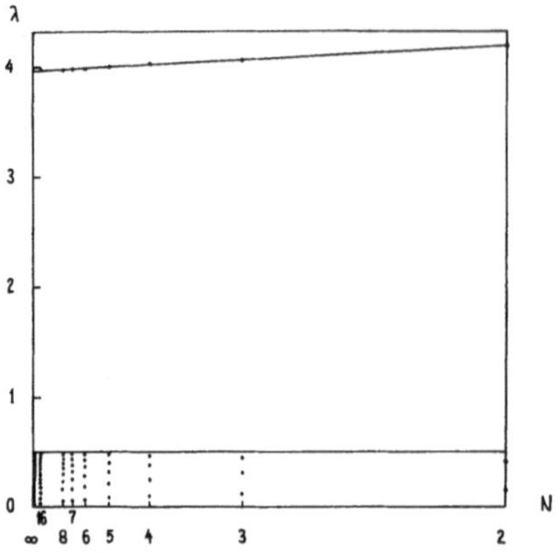

Fig. 1.13. Unpolluted numerical representation of the non-standard model problem with a continuous spectrum. As N increases, the density of modes in the region $0<\lambda<0.5$ increases

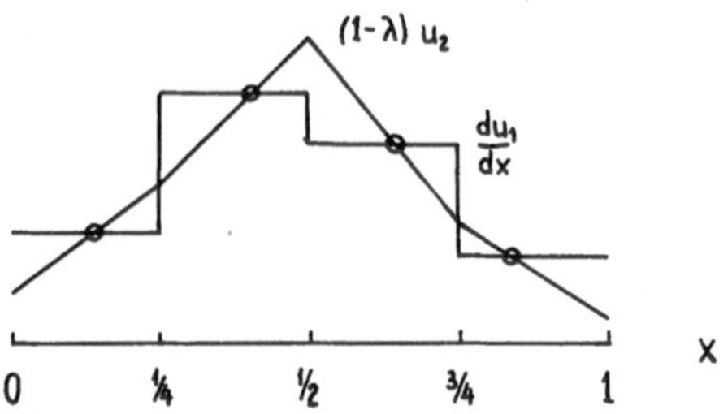

Fig. 1.14. Heuristic explanation of spectral pollution: The condition $du_1/dx-(1-\lambda)u_2=0$ can only be fulfilled at one point (0) per interval if linear elements are taken for u_1 and u_2

the case of the infinitely degenerate solution:

$$\begin{aligned}&\lambda=0\ ,\\&du_1/dx+u_2=0\ .\end{aligned}\tag{1.41}$$

Choosing hat finite elements $e_i(x)$ for the discretization of u_1 leads to piecewise constant derivatives du_1/dx. To fulfil the upper equation (1.41) everywhere, in each interval, we have to choose for u_2 the same functional dependency as for du_1/dx. This means that u_2 has to be expanded into a set of piecewise constant basis functions $c_{i-1/2}$. If we choose hat finite elements for u_2 as we did for u_1, the condition (1.41) can only be fulfilled at one single point of each interval (Fig. 1.14). This leads to the described "spectral pollution."

1.2.4 Non-Conforming Approximation

Consider the space V defined by the set of functions $\boldsymbol{u}=(u_1, u_2, u_3)$ with

$$u_1\in L^2(0,1)\,,\quad \frac{du_1}{dx}\in L^2(0,1)\,,\quad u_1(0)=0\ ,\tag{1.42}$$

$$u_2\in L^2(0,1)\ ,\tag{1.43}$$

$$\begin{aligned}&u_3\in L^2(0,1),\quad \text{such that}\quad \int_0^1(u_1-u_3)\omega\,dx=0\\&\qquad\qquad\quad \text{for all}\qquad \omega\in L^2(0,1)\ .\end{aligned}\tag{1.44}$$

The integral relation in (1.44) implies that $u_3 = u_1$. If we choose $\boldsymbol{v} = (v_1, v_2, v_3)$ in V, the bilinear forms (1.34) can be written:

$$a(\boldsymbol{u}, \boldsymbol{v}) = \int_0^1 \left(\alpha \frac{du_1}{dx} \frac{dv_1}{dx} + \beta u_2 \frac{dv_1}{dx} + \gamma u_3 v_3 + \beta \frac{du_1}{dx} v_2 + \delta u_2 v_2 \right) dx \ , \tag{1.45}$$

$$b(\boldsymbol{u}, \boldsymbol{v}) = \int_0^1 \varrho (u_3 v_3 + u_2 v_2)\, dx \ .$$

Problem (1.35) can now be reformulated:

"Find real numbers λ and non-trivial functions $\boldsymbol{u} = (u_1, u_2, u_3) \in V$ such that $a(\boldsymbol{u}, \boldsymbol{v}) = \lambda b(\boldsymbol{u}, \boldsymbol{v})$, for all $\boldsymbol{v} = (v_1, v_2, v_3) \in V$." (1.46)

We introduce a finite-dimensional space V_h to be the set of all functions $\boldsymbol{u}_h = (u_{1h}, u_{2h}, u_{3h})$ of the form

$$u_{1h} = \sum_{i=1}^{N} p_i e_i(x) \ , \tag{1.47}$$

$$u_{2h} = \sum_{i=1}^{N} q_{i-1/2} c_{i-1/2}(x) \ , \tag{1.48}$$

$$u_{3h} = \sum_{i=1}^{N} r_{i-1/2} c_{i-1/2}(x), \quad \text{such that}$$

$$\int_0^1 (u_{1h} - u_{3h}) c_{i-1/2}(x) dx = 0 \, , \quad \text{for all} \quad i = 1, \dots, N \ . \tag{1.49}$$

The functions e_i are given by (1.18, 19), and $c_{i-1/2}$ by (1.39). The p_i, $q_{i-1/2}$ and the $r_{i-1/2}$ are real numbers describing the function values of $u_{1h}(x_i)$, $u_{2h}(x_{i-1/2})$ and $u_{3h}(x_{i-1/2})$, respectively. The approximated problem is

"Find real numbers λ_h and non-trivial $\boldsymbol{u}_h = (u_{1h}, u_{2h}, u_{3h}) \in V_h$ such that $a(\boldsymbol{u}_h, \boldsymbol{v}_h) = \lambda_h b(\boldsymbol{u}_h, \boldsymbol{v}_h)$, for all $\boldsymbol{v}_h = (v_{1h}, v_{2h}, v_{3h}) \in V_h$." (1.50)

Note that the integral relation in (1.49) does not imply $u_{1h} = u_{3h}$ as is demanded in the exact problem, (1.44). The first vector component u_{1h} of $\boldsymbol{u}_h$ is a continuous function and u_{3h} is a discontinuous function. Consequently, *V_h is not included in V* and the formulation, (1.50) *is a non-conforming approximation of* (1.46).

The integral condition in (1.49) implies the following relationships between the coefficients p_i and $r_{i-1/2}$:

$$\begin{aligned} r_{1/2} &= \frac{p_1}{2} \ , \\ r_{i-1/2} &= \frac{p_{i-1} + p_i}{2}, \qquad i = 2, 3, \dots, N \ . \end{aligned} \tag{1.51}$$

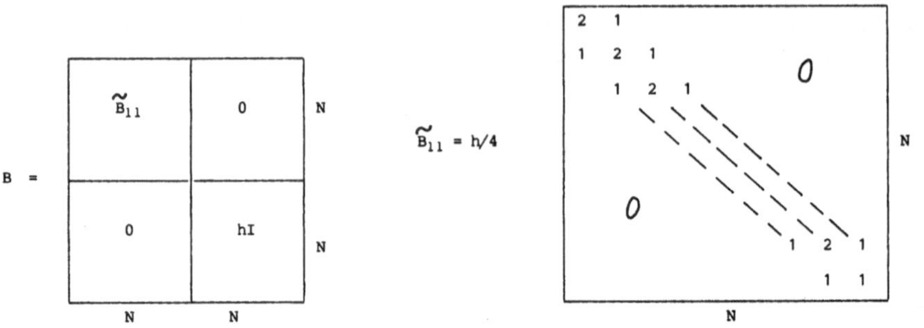

Fig. 1.15. Matrix B of the non-standard model problem when an unpolluted non-conforming finite element approach is used. Matrix A is the same as that in Fig. 1.10

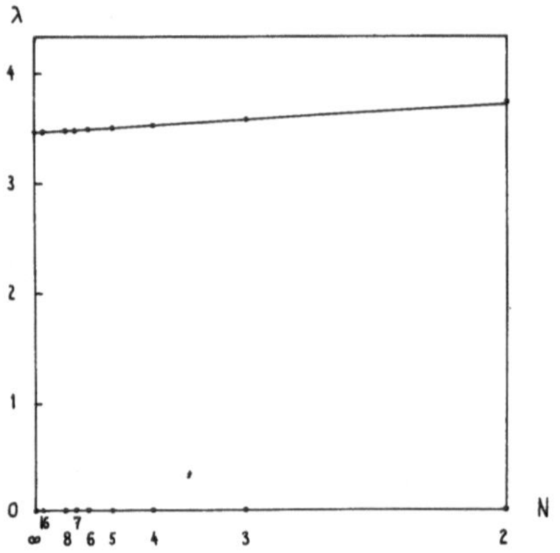

Fig. 1.16. Unpolluted numerical representation of the non-standard model problem using non-conforming finite elements for u_1 and u_2

The space V_h becomes a set of functions

$$u_h = \left(\sum_{i=1}^{N} p_i e_i,\ \sum_{i=1}^{N} q_{i-1/2} c_{i-1/2},\ \frac{p_i}{2} c_{1/2} + \sum_{i=2}^{N} \frac{p_{i-1}+p_i}{2} c_{i-1/2} \right) . \tag{1.52}$$

We apply this non-conformal approach to the first non-standard model problem, (1.25, 26). It leads to the same matrix A as for the conformal unpolluted case represented in Fig. 1.10. The matrix B is given in Fig. 1.15. The eigenvalues as a function of the mesh size h are plotted in Fig. 1.16. Like the approximate spectrum represented in Fig. 1.11, the approximate spectrum in Fig. 1.16 is non-polluted as well.

The difference between the conforming B (Fig. 1.10) and the non-conforming B (Fig. 1.15) is a result of a different integration of the first term in $b(\boldsymbol{u}, \boldsymbol{v})$. In the case of conforming elements the matrix elements reflect a Simpson integration. In the non-conforming case a "trapezoidal rule" has been used.

Note that this non-conforming finite element approach is equivalent to a direct attack of the variational form (1.35) by centered finite differences.

Advantages of the Non-Conforming Elements

A1) If hat elements are used for u_1 and piecewise constant ones for u_2 and u_3, all of the integrals can be written

$$\int_{x_{i-1}}^{x_i} f(x)dx=(x_i-x_{i-1})f(x_{i-1/2}) \ . \tag{1.53}$$

A2) The Rayleigh quotient (1.13) can reach lower values due to the extension from two (u_1, u_2) to three variables (u_1, u_2, u_3) by relaxing the condition $u_3 = u_1$, using (1.49) instead. We shall see that this relaxation will improve precision of the unstable MHD modes in the two-dimensional case.

1.3 Spectral Stability

1.3.1 General Considerations

All the problems we treat in this book can be formulated as follows:

"Find real numbers λ and non-trivial u in a Hilbert space V satisfying $a(u,v)=\lambda b(u,v)$ for all $v \in V$", (1.54)

where $a(.,.)$ and $b(.,.)$ are bilinear forms defined on V. Denoting by $(.,.)$ the scalar product in V and by $\|\cdot\|$ the norm, i.e. $\|w\|=(w,w)^{1/2}$ if $w \in V$, we shall assume the following hypotheses on $a(.,.)$ and $b(.,.)$:

H1) $a(.,.)$ and $b(.,.)$ are *symmetric forms*, i.e. $a(u,v)=a(v,u)$ and $b(u,v) = b(v,u)$ for all $u, v \in V$.

H2) $a(.,.)$ and $b(.,.)$ are *continuous* in V, i.e. there exists a positive constant c such that

$$|a(u,v)| \leqq c\|u\| \cdot \|v\| \quad \text{and}$$

$$|b(u,v)| \leqq c\|u\| \cdot \|v\| \quad \text{for all} \quad u, v \in V \ .$$

H3) $a(.,.)$ is *V-elliptic* or *coercive* on V, i.e. there exists a positive constant α such that $a(u,u) \geqq \alpha\|u\|^2$ for all $u \in V$. In practice this hypothesis on $a(.,.)$ is satisfied through an eigenvalue shift.

For the standard model example (1.11), (H1) and (H2) are trivially satisfied when we choose V as being the set of all functions $w \in L^2(0,1)$ with $dw/dx \in L^2(0,1)$ and $w(0)=0$, with the scalar product

$$(u,v)=\int_0^1 \left(\frac{du}{dx}\frac{dv}{dx}+uv\right)dx \ . \tag{1.55}$$

By adding $\Lambda_0 \cdot b(u, v)$ on both sides of (1.12), where the eigenvalue shift Λ_0 is a real positive constant, we can satisfy (H3) by choosing Λ_0 big enough, provided that $\varrho > 0$.

The same considerations can be made for the non-standard problem (1.33) with the Hilbert space V defined by all functions $\boldsymbol{u} = (u_1, u_2)$ with $u_1 \in L^2(0, 1)$, $du_1/dx \in L^2(0, 1)$, $u_1(0) = 0$, $u_2 \in L^2(0, 1)$ and the scalar product

$$(u, v) = \int_0^1 \left(\frac{du_1}{dx} \frac{dv_1}{dx} + u_1 v_1 + u_2 v_2 \right) dx \ . \tag{1.56}$$

If $a(.,.)$ fulfils (H1) to (H3), $a(.,.)$ is a scalar product on V, equivalent to $(.,.)$[8]. Then the problem.

$$\begin{aligned} &\text{“Given } f \in V, \text{ find } w \in V \quad \text{such that} \\ &a(w, v) = b(f, v) \quad \text{for all} \quad v \in V\text{”}, \end{aligned} \tag{1.57}$$

has a unique solution w (Lax-Milgram theorem) (*Agmon* 1965). If we call A the linear operator on V which maps f to w, then (1.57) can be written

$$a(Af, v) = b(f, v) \quad \text{for all} \quad f \quad \text{and} \quad v \in V \ . \tag{1.58}$$

It can be seen that A is a self-adjoint[9] continuous[10] linear operator on V, if V is provided with the scalar product $a(.,.)$. If (λ, u) is an eigenpair of (1.54), we have, by using (1.58),

$$a(u, v) = \lambda b(u, v) = \lambda a(Au, v) \quad \text{for all} \quad v \in V \ . \tag{1.59}$$

Consequently, $a(u - \lambda Au, v) = 0$ for all $v \in V$, which implies that $u - \lambda Au = 0$. One can easily see that (λ, u) is an eigenpair of (1.54) if, and only if, $\mu = 1/\lambda$ is an eigenvalue of A corresponding to the eigenvector u. Therefore, solving (1.54) is equivalent to the problem of finding the eigenelements of the operator A.

Let us recall some standard definitions:

The *resolvent set* of the operator A is the set of $\mu \in R$ such that $(\mu I - A)$ is continuously invertible on V, where I denotes the identity operator of V.[11] The spectrum $\sigma(A)$ of A is the set of all the values μ which does not belong to the resolvent set of A. A value $\mu \in \sigma(A)$ is called an isolated eigenvalue with multiplicity m if μ is an isolated point of $\sigma(A)$, and if the eigenspace of A corresponding to μ is of dimension m. The continuous spectrum of A is the set of values $\mu \in \sigma(A)$ which are not eigenvalues of A.[12]

---

8 There exist positive constants c_1 and c_2 such that
$c_1 \|u\|^2 \leqq a(u, u) \leqq c_2 \|u\|^2$ for all $u \in V$

9 $a(Af, v) = a(f, Av)$ for all $f, v \in V$

10 $\sup\limits_{f \in V, f \neq 0} \dfrac{\|Af\|}{\|f\|} < \infty$

11 Since A is self-adjoint, we only consider real values for μ

12 This definition holds since A is self-adjoint

For the classical problem, (1.12), one can prove that, if $\mu \in \sigma(A)$, $\mu \neq 0$, then μ is an isolated eigenvalue with multiplity 1. The non-standard problem (1.35), however, gives rise to a spectrum $\sigma(A)$ which may contain some values μ which are not eigenvalues with finite multiplicity.[13] Since the continuous spectrum is part of the physical problem, we will be interested by the spectrum $\sigma(A)$ rather than by the eigenvalues of A.

Let us now consider an approximation of (1.54) using a Galerkin method. We denote by V_h a finite-dimensional subspace of V (in practice h is the mesh-size parameter which tends to zero) and suppose that

H4) $\lim_{h\to 0} \min_{u_h\in V_h} \|u-u_h\|=0, \quad \text{for all} \quad u\in V\,.$ (1.60)

Then, the numerical approximation of (1.54) consists in replacing V by V_h, giving the problem

"Find $\lambda_h \in R$, $u_h \in V_h$, $u_h \neq 0$, such that
$a(u_h, v_h) = \lambda_h b(u_h, v_h)$, for all $v_h \in V_h$." (1.61)

The approximation of the classical problem by linear finite elements satisfies (H4). Both the "polluting" and the "non-polluting" approximations of the non-standard problem satisfy (H4).

Under hypotheses (H1–3) we can define the linear operator A_h from V_h into itself, which approximates A, by

$$a(A_h f, v_h) = b(f, v_h) \quad \text{for all} \quad f, v_h \in V_h\,. \tag{1.62}$$

Again, (λ_h, u_h) is an eigenpair of (1.61) if, and only if, $\mu_h = 1/\lambda_h$ is an eigenvalue of A_h corresponding to the eigenvector u_h. Since A_h is an operator defined on a finite-dimensional space V_h, the spectrum $\sigma(A_h)$ of A_h contains only eigenvalues of finite multiplicity. Moreover, A_h is a self-adjoint operator on V_h if V_h is provided with the scalar product $a(.,.)$. The question which now arises is, how is the spectrum of A approximated by $\sigma(A_h)$?

1.3.2 Stability Conditions

The questions (I) to (III) we asked in Sect. 1.1.3 can be answered positively if the following properties are fulfilled:

I) For $\mu \in \sigma(A)$, there exists $\mu_h \in \sigma(A_h)$ with $\lim_{h\to 0} |\mu - \mu_h| = 0$.

An isolated eigenvalue of A of finite multiplicity is approximated by eigenvalues of A_h of the same total multiplicity, if $h \leq h_0$ is small enough.

II) If κ is a closed subset of R which does not intersect $\sigma(A)$, then κ does not intersect $\sigma(A_h)$ if $h \leq h_0$ is small enough.

13 Mathematically, the classical problem gives rise to a compact operator A, whereas in the non-standard problem A is not compact

III) Invariant subspaces of A are approximated by corresponding invariant subspaces of A_h when h tends to zero.[14]

For the classical problem, (1.12), hypothesis (H4), (1.60) is sufficient to obtain the properties (I–III). However, in more general cases such as our non-standard problem, (1.35), hypothesis (H4) is not sufficient to satisfy (II). We then have a sequence $\mu_h \in \sigma(A_h)$ such that $\lim_{h\to 0} \mu_h = \mu$, which does not belong to $\sigma(A)$ (*Rappaz* 1976, 1977).

We propose to call this a "polluting" approximation (Fig. 1.8). To suppress pollution, an additional numerical stability condition for the spectral approximation has to be included:

H5) $$\lim_{h\to 0} \sup_{\substack{u_h \in V_h \\ \|u_h\| = 1}} \|Au_h - A_h u_h\| = 0 \ .$$

This leads us to the main theorem (*Descloux* et al. 1978a):

Theorem 1. Hypotheses (H4) and (H5) imply that properties (I–III) are satisfied. This is the "non-pollution" theorem. It tells us that when construction subspaces V_h of V which satisfy (H4) and (H5), the properties (I–III) are fulfilled. For the approximation of the non-standard problem, one can prove that the space V_h spanned by the basis functions, (1.40), satisfies (H4) and (H5) (*Descloux* et al. 1977).

Remark. So far we have discussed only the conforming approximation of Problem (1.54). In the non-conforming case, the situation is slightly different since V_h is not included in V. We can introduce a larger space W such that V_h and V are two subspaces of W. The operators A_h and A admit continuous extensions on W, i.e. A can be considered as an operator $\tilde{A}$ from W into V and A_h as an operator $\tilde{A}_h$ from W into V_h, such that the restriction of $\tilde{A}$ to V is equal to A, and the restriction of $\tilde{A}_h$ to V_h is equal to A_h. With this notation, one can prove that our main Theorem (1) holds if A is replaced by $\tilde{A}$ in (H5) (*Evequoz* 1980). The non-conforming approach of subsection 1.2.4 satisfies this hypothesis and consequently is a "non-polluting" approach.

1.3.3 Order of Convergence

We intend to give some simple considerations which concern the rate of convergence of symmetric eigenproblems. We assume (H1) to (H3) and let $u \in V$, $\lambda \in R$, $u_h \in V_h$, $\lambda_h \in R$ be such that

$$a(u, v) = \lambda b(u, v), \quad \text{for all} \quad v \in V \ , \tag{1.63}$$

$$a(u_h, v_h) = \lambda_h b(u_h, v_h), \quad \text{for all} \quad v_h \in V_h \ , \tag{1.64}$$

14 For an exact formulation of this property, see *Descloux* et al. (1978a). In particular, if μ is an isolated eigenvalue with the finite multiplicity, we want the corresponding eigensubspace to be approximated by the space spanned by the eigenvectors of A_h, which correspond to the eigenvalues μ_h with $\lim_{h\to 0} \mu_h = \mu$

with the normalizations

$$a(u,u)=1 \ , \tag{1.65}$$

$$a(u_h,u_h)=1 \ . \tag{1.66}$$

Since λ and λ_h cannot be zero using (1.63–66), we have

$$\begin{aligned}\frac{1}{\lambda_h}-\frac{1}{\lambda}&=b(u_h,u_h)-\frac{1}{\lambda}a(u_h,u_h)\\&=b(u_h-u,u_h-u)-\frac{1}{\lambda}a(u_h-u,u_h-u) \ .\end{aligned} \tag{1.67}$$

Moreover, we have $|\lambda-\lambda_h|=|\lambda\cdot\lambda_h|\left|\frac{1}{\lambda_h}-\frac{1}{\lambda}\right|$.

By using (H2) and (1.67) we obtain

Theorem 2. If $\lim_{h\to 0}\lambda_h=\lambda$ *and* $\lim_{h\to 0}u_h=u\in V$, there exists a constant C independent of h such that $|\lambda-\lambda_h|\leqq C\|u-u_h\|^2$.

This Theorem (2) can be found in *Descloux* (1979).

In many practical cases, specifically in the finite element method, we can give error estimates for $\|u-u_h\|$ or, most generally, error estimates between the invariant subspaces of A and the corresponding invariant subspaces of A_h (*Descloux* et al. 1978b). All "non-polluting" conforming and non-conforming approximations with linear and piecewise constant finite elements of problem (1.22) lead to error estimates of the type[15]

$$|\lambda-\lambda_h|=0(h^2) \ . \tag{1.68}$$

1.4 Finite Elements of Order p

In this section we very briefly present the finite element method of order p to approximate the problems of Sects. 1.1 and 1.2. For this purpose we subdivide the interval $[0,1]$ into N parts $0=x_0<x_1<x_2<\ldots<x_N=1$, where the x_i are real numbers which we call nodes.

1.4.1 Discontinuous Finite Elements S_p^0

If p is a non-negative integer, we denote by S_p^0 the set of all the functions in $L^2(0,1)$ whose restriction to each interval $(x_{i-1},\ x_i)$, $i=1,2,\ldots,N$ is a polynomial of degree p.

15 $|\lambda-\lambda_h|=0(h^2)$ means there exists a constant C (independent of h) such that $|\lambda-\lambda_h|\leqq Ch^2$

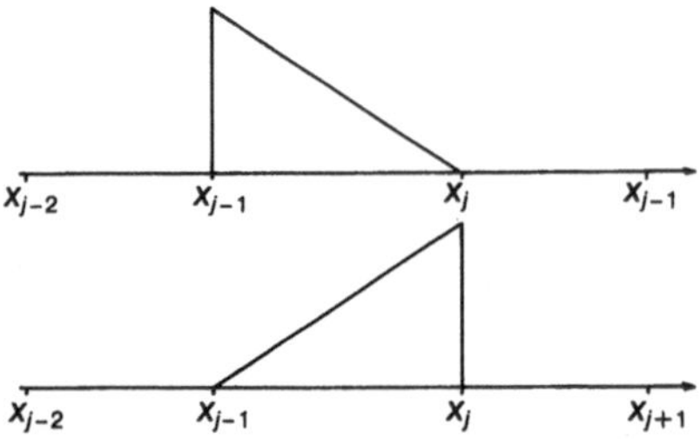

Fig. 1.17. Discontinuous linear basis functions of S_1^0 in the interval $x_{j-1} \leq x \leq x_j$

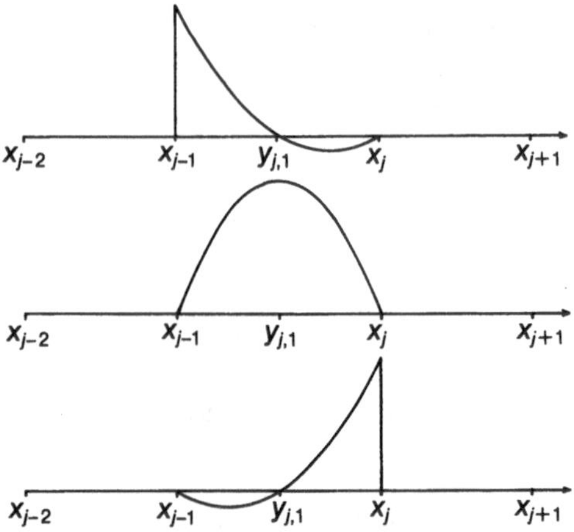

Fig. 1.18. Discontinuous quadratic basis functions of S_2^0 which are non-zero in the interval $x_{j-1} \leq x \leq x_j$

For $p=0$, the set S_0^0 of piecewise constant basis functions $c_{i-1/2}$ were given in (1.39) and in Fig. 1.9. In order to construct a basis of S_p^0 for $p \geqq 1$, we have to choose $p-1$ interior nodes in the interval (x_{i-1}, x_i) denoted by

$$x_{i-1} \equiv y_{i,0} < y_{i,1} < y_{i,2} < \ldots < y_{i,p-1} < y_{i,p} \equiv x_i \ . \tag{1.69}$$

The set S_p^0 can be constructed from a linear combination of $N(p+1)$ basis functions $\Phi_{i,j}(x)$, $j=0, 1, \ldots, p$, $i=1, 2, \ldots, N$ defined by

$$\begin{aligned} \Phi_{i,j}(x) &= 0 && \text{if} \quad x \notin [x_{i-1}, x_i] \ , \\ \Phi_{i,j}(x) &= \frac{\prod_k' (x - y_{i,k})}{\prod_k' (y_{i,j} - y_{i,k})} && \text{if} \quad x \in (x_{i-1}, x_i) \ , \end{aligned} \tag{1.70}$$

where the notation $\prod_k'$ means $\prod_{k=0, k \neq j}^{p}$.

We remark that $\Phi_{i,j}(y_{m,n}) = \delta_{im}\delta_{jn}$ and functions in S_p^0 jump at the nodal points x_i, $i=0, 1, \ldots, N$. The $2N$ basis functions of S_1^0 are shown in Fig. 1.17 and the $3N$ basis functions of S_2^0 in Fig. 1.18.

1.4.2 Continuous Finite Elements S_p^1 (Lagrange Elements)

If p is an integer with $p \geqq 1$, we denote by S_p^1 the set of all *continuous functions* defined on $[0, 1]$ whose restriction to each interval $[x_{i-1}, x_i]$, $i=1, 2, \ldots, N$, is a polynomial of degree p.

For $p=1$, the set S_1^1 of linear basis functions $e_i(x)$ were given in (1.18, 19, 37) and Fig. 1.2. In order to construct a basis of S_p^1 for $p \geqq 2$ we have to choose $p-1$ interior nodes in each interval (x_{i-1}, x_i) [see (1.69)].

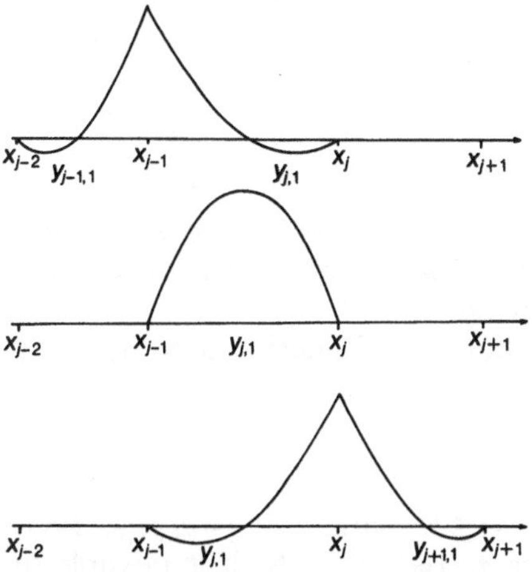

Fig. 1.19. Continuous quadratic basis functions of S_2^1 which are non-zero in the interval $x_{j-1} \leq x \leq x_j$

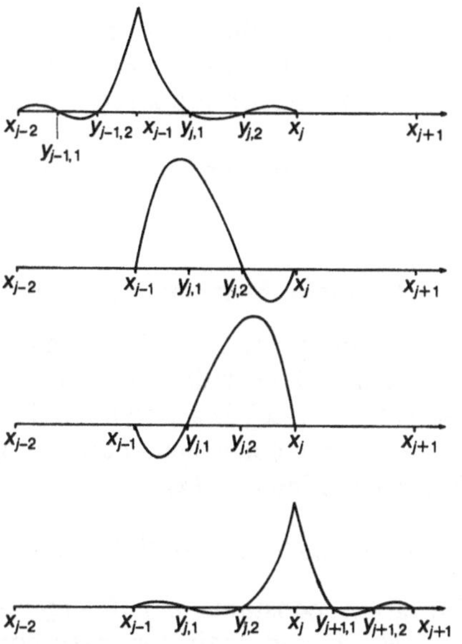

Fig. 1.20. Continuous cubic Lagrange basis functions of S_3^1 which are non-zero in the interval $x_{j-1} \leq x \leq x_j$

The set S_p^1 can be constructed from a linear combination of $N(p-1)$ basis functions $\Phi_{i,j}(x)$, $j=1,2,\ldots,p-1$, $i=1,2,\ldots,N$ with $(N+1)$ basis functions $\Phi_i(x)$, $i=0,1,2,\ldots,N$ defined by

$$
\begin{aligned}
&\Phi_{i,j}(x)=0 && \text{if} \quad x \notin [x_{i-1}, x_i] \ , \\
&\Phi_{i,j}(x)=\frac{\prod_k' (x-y_{i,k})}{\prod_k' (y_{i,j}-y_{i,k})} && \text{if} \quad x \in [x_{i-1}, x_i] \ ,
\end{aligned}
\tag{1.71}
$$

and by setting $x_{-1}=0$, $x_{N+1}=1$,

$$
\begin{aligned}
&\Phi_i(x)=0 && \text{if} \quad x \notin [x_{i-1}, x_{i+1}] \ , \\
&\Phi_i(x)=\frac{\prod_{k=0}^{p-1} (x-y_{i,k})}{\prod_{k=0}^{p-1} (x_i-y_{i,k})} && \text{if} \quad x \in [x_{i-1}, x_i], \\
&\Phi_i(x)=\frac{\prod_{k=1}^{p} (x-y_{i+1,k})}{\prod_{k=1}^{p} (x_i-y_{i+1,k})} && \text{if} \quad x \in [x_i, x_{i+1}] \ .
\end{aligned}
\tag{1.72}
$$

We remark that the functions $\Phi_{i,j}$ and Φ_i are continuous functions on $[0,1]$ with the value 1 at an interior node and zero at the other nodes. Moreover, the functions $\Phi_{i,j}$ have a support of 1 interval and the functions Φ_i have a support of 2 intervals.

The basis functions of S_2^1 are shown in Fig. 1.19 and the basis functions of S_3^1 in Fig. 1.20.

1.4.3 C^1-Finite Elements S_p^2 (Hermite Elements)

If p is an integer with $p \geqq 3$, we denote by S_p^2 the set of all *continuous functions* defined on $[0,1]$ with *continuous first derivative*, whose restriction to each interval $[x_{i-1}, x_i]$ is a polynomial of degree p.

In order to construct a basis for S_p^2 we choose $p-3$ interior nodes $y_{i,k}$ such that in each interval (x_{i-1}, x_i)

$$x_{i-1} \equiv y_{i,0} < y_{i,1} < \ldots < y_{i,p-3} < y_{i,p-2} \equiv x_i \ . \tag{1.73}$$

For given constants $c_{i,j}$, $0 \leqq j \leqq p-2$, $1 \leqq i \leqq N$ with $c_{i,p-2} = c_{i+1,0}$, $i=1, 2, \ldots, N-1$, and again for given constants d_i, $i=0,1,\ldots,N$, it is possible to construct a unique C^1 function h defined on $[0,1]$, the restriction of which to each interval $[x_{i-1}, x_i]$ is a polynomial of degree p; h verifies:

$$\begin{aligned} &h(y_{i,j}) = c_{i,j}, \quad j=0,1,\ldots,p-2, \quad i=1,2,\ldots,N \quad \text{and} \\ &\frac{dh}{dx}(x_i) = d_i, \quad i=0,1,\ldots,N \ . \end{aligned} \tag{1.74}$$

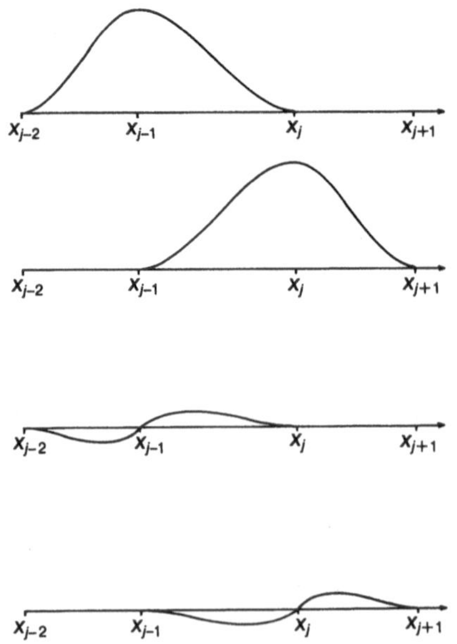

Fig. 1.21. Continuous cubic Hermite basis functions of S_3^2 with continuous first derivatives which are non-zero in the interval $x_{j-1} \leq x \leq x_j$

A basis function of S_p^2 is obtained if we fix one coefficient $c_{i,j}$ or d_i to 1 and all the others to zero. The basis functions of S_3^2 are (see Fig. 1.21.):

$$h_i(x)=\begin{cases}\dfrac{(x-x_{i-1})^2(3x_i-x_{i-1}-2x)}{(x_i-x_{i-1})^3} & \text{if} \quad x\in[x_{i-1},x_i] \ ,\\ \dfrac{(x-x_{i+1})^2(3x_i-x_{i+1}-2x)}{(x_i-x_{i+1})^3} & \text{if} \quad x\in[x_i,x_{i+1}] \ ,\\ 0 & \text{if} \quad x\notin[x_{i-1},x_{i+1}] \ ,\end{cases} \tag{1.75}$$

and

$$h_i^*(x)=\begin{cases}\dfrac{(x-x_{i-1})^2(x-x_i)}{(x_i-x_{i-1})^2} & \text{if} \quad x\in[x_{i-1},x_i] \ ,\\ \dfrac{(x-x_{i+1})^2(x-x_i)}{(x_i-x_{i+1})^2} & \text{if} \quad x\in[x_i,x_{i+1}] \ ,\\ 0 & \text{if} \quad x\notin[x_{i-1},x_{i+1}] \ .\end{cases} \tag{1.76}$$

1.4.4 Application to the Model Problems

Let us apply the conforming elements to the classical model problem (1.1). We assume that the coefficients α, β and ϱ are infinitely continuously differentiable functions. Taking for the space $V_h=\{f\in S_p^1 \text{ with } f(0)=0\}$, it is well known that the error estimate for the eigenvalues is (*Babuska* and *Aziz* 1972)

$$|\lambda-\lambda_h|=0(h^{2p}) \ . \tag{1.77}$$

We now apply the conforming finite elements to the *non-standard problem* (1.22). Again, we assume that the coefficients α, β, γ, ϱ and δ are infinitely continuously differentiable functions. For $p\geqq 1$, we define V_h to be the set of all functions $\boldsymbol{u}=(u_1, u_2)$, such that $u_1\in S_p^1$, $u_1(0)=0$ and $u_2\in S_{p-1}^0$. Choosing such a space V_h, one can prove (*Descloux* et al. 1977) that the stability conditions given in Sect. 1.3.2 are satisfied. Moreover, these spaces lead to error estimates of (*Descloux* et al. 1977, 1978b)

$$|\lambda-\lambda_h|=0(h^{2p}) \ . \tag{1.78}$$

The same convergence properties are found for $p\geqq 3$ when we choose for $V_h=\{\boldsymbol{u}=(u_1, u_2)\colon u_1\in S_p^2,\ u_1(0)=0 \text{ and } u_2\in S_{p-1}^1\}$. The proof of this assertion can be established by the same arguments as those given in *Descloux* et al. (1977).

1.4.5 Non-Conforming Lagrange Elements

In basis functions of the non-conforming approach (1.50) the component u_{1h} has to be taken in S_p^1 where $p\geqq 1$ and u_{2h} and u_{3h} in S_{p-1}^0. In (1.51, 52) the particular choice of $p=1$ was chosen. Here, as a result of the integral condition in (1.49), the non-conforming approach is identical to a direct attack of the variational formulation by finite differences. We can generalize this approach to $p \geqq 1$.

We take for u_{1h}, *continuous functions* S_p^1 *which are polynomials of order p in each interval*. Consequently, for u_{2h} and u_{3h}, one has to choose *discontinuous functions* S_{p-1}^0 *(see Figs. 1.17 and 18)* which are polynomials of order $p-1$ in each interval. For $p=2$ there are three nodal points of u_{1h} and two nodal points of u_{2h} and u_{3h} in each interval. We now have to identify the two nodal values of u_{3h} with the three nodal values of u_{1h} through the integral condition (1.44), taking as a test function ω the two discontinuous linear functions shown in Fig. 1.17. Denoting by p_{i0}, p_{i1} and p_{i2} the three values of u_{1h} at the nodes x_i, $x_{i+1/2}$, x_{i+1}, and by r_{i0} and r_{i1} the two values of u_{3h}, at the nodes x_i, x_{i+1} the relations are

$$\begin{aligned} r_{i0} &= (2p_{i0}+2p_{i1}-p_{i2})/3 \ , \\ r_{i1} &= (2p_{i2}+2p_{i1}-p_{i0})/3 \ . \end{aligned} \tag{1.79}$$

1.4.6 Non-Conforming Hermite Elements with Collocation

The choice of Lagrange elements for u_{1h} has the advantage that the integral conditions (1.44) lead to block diagonal systems of linear equations which can be solved explicitly in each interval. This is due to the discontinuous behavior of u_{3h}.

Using now Hermite elements S_p^2 for u_{1h} with continuous first derivatives leads us to choose continuous functions S_{p-1}^1 for u_{3h}. If we now take in the integral condition (1.44) functions ω of the type S_{p-1}^1, then the unknowns in between intervals are coupled, and it is not possible to express the unknowns corresponding to u_{3h} as functions of the unknowns corresponding to u_{1h} explicitly in each interval of the discretization.

To overcome this disadvantage we propose a *heuristic solution*. Impose continuity on u_{3h} through collocation and impose the integral condition (1.44) only with a test function ω in S_{p-1}^1 corresponding to internal points of the element.

Let us apply this heuristic approach to the case $p=3$. The 4 unknowns of u_{1h} are denoted by p_i, p_{i+1}, p_i^* and p_{i+1}^*, and the 3 unknowns of u_{3h} by r_i, $r_{i+1/2}$ and r_{i+1}. Collocation at the nodes leads to

$$\begin{aligned} r_i &= p_i \ , \\ r_{i+1} &= p_{i+1} \ . \end{aligned} \tag{1.80}$$

The condition between $r_{i+1/2}$ and the four values of u_{1h} is obtained by choosing $\Phi_{i+1,1}$ (see Fig. 1.19) as a test function in (1.44). It then reads

$$r_{i+1/2} = (p_i + p_{i+1})/2 - [(x_{i+1}-x_i)/8]\,(p_{i+1}^* - p_i^*) \ . \tag{1.81}$$

This latter condition corresponds to a collocation at the mid point of the cell.

1.5 Some Comments

It is possible to consider the two-dimensional (2D) analogon [see *Strang* and *Fix* (1973) or *Ciarlet* (1978)] of the classical one-dimensional (1D) eigenproblem treated in Sect. 1.1 and a 2D analogon of the non-standard 1D problem (see *Rappaz* 1979, *Jaccard* 1980; *Evequoz* 1980; *Evequoz* and *Jaccard* 1980; *Jaccard* and *Evequoz* 1982) discussed in Sects. 1.2–4. The mathematical analysis is similar to that used for the 1D problem, but technically more complicated.

To treat such a 2D eigenvalue problem numerically, we use classical 2D finite elements (see *Ciarlet* 1978) for the classical problem and, in the non-standard case, we have to construct finite dimensional subspaces which satisfy the stability conditions of Sect. 1.3. If these conditions are not satisfied, in principle we obtain a "polluting" approximation as shown in Sect. 1.2.2. Some mathematical information on the polluting effect can be obtained in *Descloux* (1981), *Rappaz* (1982).

For those who are interested in the error estimates of the approximation of eigenvalue problems, we refer to articles by *Bramble* and *Osborn* (1973), *Fix* (1973), *Osborn* (1975), *Descloux* et al. (1978b), and by *Descloux* (1979). As reference on the mathematical analysis of the finite element method we propose *Ciarlet* (1978). The spectral stability of finite element methods is studied by *Descloux* et al. (1979) and *Mills* (1979).

Finally, we propose the book by *Zienkiewicz* (1977) for those readers who are mainly interested in applications of the finite element method.

2. The Ideal MHD Model

2.1 Basic Equations

The plasma state is often called the fourth state following the solid, liquid and gaseous states. If a gas is heated above, say, 10 000 K, the gas is ionized due to collisions between particles. Such a mixture of ions and electrons is called a plasma (*Chen* 1974). If the temperature is raised to temperatures necessary in a thermonuclear reactor ($T \geqq 10\,\mathrm{keV} \approx 10^8\,\mathrm{K}$) almost all electrons are free and the plasma becomes a very good conductor. In order to reach such high temperatures, the ionized hot gas has to be kept away from material walls, and the plasma has to be confined. This is done by applying a strong magnetic field, which acts on the charged particles in such a way that it provides a counterpressure to the gas pressure. A plasma in a magnetic bottle behaves like a mixture of fluids which can be described by the fluid equations coupled to the Maxwell equations. If one neglects the relative motion between ions and electrons and considers the plasma as only one averaged fluid, one uses the magnetohydrodynamic (MHD) equations. In the special case of an infinitely good conducting gas (with resistivity = 0), one uses what are called the ideal MHD equations. It is exactly this most simple model that we will treat numerically in the following chapters. It describes surprisingly well the equilibrium state of a magnetically confined plasma, and the rapid unstable global motions which can destroy the confinement on a microsecond timescale.

The most general domain we intend to consider is shown in Fig. 2.1. The plasma region Ω_p is surrounded by a vacuum region Ω_v, bounded by a conducting wall Γ_v which isolates the whole system. The plasma-vacuum interface is denoted by Γ_p. In the following we denote by $\boldsymbol{n}$ the normal vector to

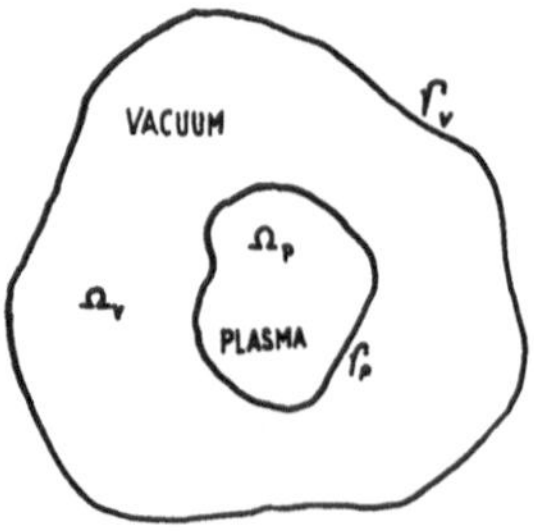

Fig. 2.1. The plasma region Ω_p bounded by Γ_p, surrounded by a vacuum region Ω_v. A conducting shell is put at Γ_v

surfaces Γ_p, Γ_v, ..., pointing outwards. The set of ideal MHD equations reads (*Kadomtsev* 1966; *Schmidt* 1966; *Bateman* 1978; *Miyamoto* 1980; *Artsimowitsch* and *Sagdejew* 1983) in Ω_p:

$$\begin{aligned}
&\frac{D\varrho}{Dt} = -\varrho\nabla\cdot\boldsymbol{v} \quad \text{(continuity)}\\
&\varrho\frac{D\boldsymbol{v}}{Dt} = -\nabla p+\boldsymbol{J}\times\boldsymbol{B} \quad \text{(Newton)}\\
&\frac{D}{Dt}(p\varrho^{-\gamma})=0 \quad \text{(equation of state)}\\
&\frac{\partial\boldsymbol{B}}{\partial t}=\nabla\times(\boldsymbol{v}\times\boldsymbol{B}) \quad \text{(Ohm's law)}\\
&\left.\begin{array}{l}\nabla\times\boldsymbol{B}=\boldsymbol{J}\\ \nabla\cdot\boldsymbol{B}=0\ ,\end{array}\right. \quad \text{(Maxwell)}
\end{aligned} \tag{2.1}$$

and in Ω_v:

$$\left.\begin{array}{l}\nabla\times\boldsymbol{B}_v=0\\ \nabla\cdot\boldsymbol{B}_v=0\end{array}\right. \quad \text{(Maxwell)} \tag{2.2}$$

subject to boundary conditions which we will discuss when we solve (2.1, 2) for a static equilibrium and, when perturbing it, for the eigensolution of the eigenvalue problem. The quantities ϱ, p, γ, $\boldsymbol{v}$, and $\boldsymbol{J}$ are the mass density, pressure, adiabaticity, velocity of the fluid and the current density, respectively. The magnetic field in Ω_p is denoted by $\boldsymbol{B}$ and by $\boldsymbol{B}_v$ in Ω_v. The natural system of units of *Weibel* (1968) is used throughout this work. This implies that the magnetic permeability $\mu_0=1$ and the velocity of light $c=1$. D/Dt denotes the convective derivative,

$$\frac{D}{Dt}\equiv\frac{\partial}{\partial t}+\boldsymbol{v}\cdot\nabla\ . \tag{2.3}$$

The pressure is assumed to be isotropic. Ohm's law follows from the assumption of infinite conductivity and can be interpreted (*Schmidt* 1966) as a freezing of the magnetic lines of force in the fluid.

2.2 Static Equilibrium

In a static equilibrium state all of the quantities in (2.1) are time independent ($\partial/\partial t=0$), and $\boldsymbol{v}=0$. Therefore, the convective derivative

$$\frac{D}{Dt}=0\ . \tag{2.4}$$

We thus exclude stationary solutions with $\boldsymbol{v} \neq 0$. Adding the index "0" to equilibrium quantities, the remaining non-trivial MHD equations are

$$\begin{aligned} \nabla p_0 &= \boldsymbol{J}_0 \times \boldsymbol{B}_0 \\ \nabla \times \boldsymbol{B}_0 &= \boldsymbol{J}_0 \\ \nabla \cdot \boldsymbol{B}_0 &= 0 \ . \end{aligned} \tag{2.5}$$

For simplicity, we always solve these equilibrium equations with the fixed boundary conditions, implying that

$$\boldsymbol{B}_0 \cdot \boldsymbol{n} = 0 \quad \text{at} \quad \Gamma_\mathrm{p} = \Gamma_\mathrm{v} \ . \tag{2.6}$$

A confined plasma in equilibrium must have closed isobars, which do not intersect any material surface. Dotting the first equation in (2.5) with $\boldsymbol{B}_0$ and $\boldsymbol{J}_0$, successively, shows that isobaric surfaces are also magnetic and current surfaces.

Equations (2.5) do not allow a confinement configuration topologically equivalent to a sphere. The simplest possible closed configurations are toroidal. Another two-dimensional configuration is a straight helix as it appears in the case of a stellerator with infinite aspect ratio. A special case of this class of equilibria is the one-dimensional, straight, infinitely long cylinder. Note that mirror configurations are based upon anisotropic pressure effects and cannot be described by (2.5). The plasma boundary has to be an isobar if there is a vacuum around the plasma. Except, where mentioned explicitly, we shall always assume that the current density $\boldsymbol{J}_0$ is finite in Ω_p and zero in Ω_v which implies no surface currents, that $p_0(\Gamma_\mathrm{p}) = 0$, and that $\boldsymbol{B}_0$ is continuous across the boundary. As far as the plasma is concerned, the equilibrium solution only depends on $\boldsymbol{B}_0$ at the surface, and not on $\boldsymbol{B}_0$ in the vacuum region.

2.3 Linearized MHD Equations

Let us assume a plasma in equilibrium described by (2.5). If at time $t = 0$ the system is slightly displaced from its equilibrium, it will, in general, evolve in time. As long as the displacement is small, we can study this evolution by linearizing (2.1, 2) around the equilibrium solution. Defining by $\delta\varrho$, δp, $\delta\boldsymbol{B}$, $\delta\boldsymbol{J}$, and $\boldsymbol{v}$ the perturbations in the corresponding quantities, the total values of the quantities are:

$$\left.\begin{aligned} \boldsymbol{v}(\boldsymbol{r},t) &= \boldsymbol{0} + \boldsymbol{v}(\boldsymbol{r},t) \\ \varrho(\boldsymbol{r},t) &= \varrho_0(\boldsymbol{r}_0) + \delta\varrho(\boldsymbol{r},t) \\ p(\boldsymbol{r},t) &= p_0(\boldsymbol{r}_0) + \delta p(\boldsymbol{r},t) \\ \boldsymbol{B}(\boldsymbol{r},t) &= \boldsymbol{B}_0(\boldsymbol{r}_0) + \delta\boldsymbol{B}(\boldsymbol{r},t) \\ \boldsymbol{J}(\boldsymbol{r},t) &= \boldsymbol{J}_0(\boldsymbol{r}_0) + \delta\boldsymbol{J}(\boldsymbol{r},t) \end{aligned}\right\} \quad \text{in} \quad \Omega_\mathrm{p} \ . \tag{2.7}$$

$$\boldsymbol{B}_\mathrm{v}(\boldsymbol{r},t) = \boldsymbol{B}_{\mathrm{v}0}(\boldsymbol{r}_0) + \delta\boldsymbol{B}_\mathrm{v}(\boldsymbol{r},t) \quad \text{in} \quad \Omega_\mathrm{v} \ . \tag{2.8}$$

Now we can insert these quantities in (2.1), keeping only the linear terms in the perturbed quantities. After the elimination of δp and $\delta \boldsymbol{J}$, we obtain

$$\begin{aligned}
&\varrho_0 \frac{\partial \boldsymbol{v}}{\partial t} = -\nabla \delta p + (\nabla \times \delta \boldsymbol{B}) \times \boldsymbol{B}_0 + \boldsymbol{J}_0 \times \delta \boldsymbol{B} \\
&\frac{\partial \delta p}{\partial t} = -\boldsymbol{v} \cdot \nabla p_0 - \gamma p_0 \nabla \cdot \boldsymbol{v} \qquad \text{in} \quad \Omega_{\mathrm{p}} \ . \\
&\frac{\partial \delta \boldsymbol{B}}{\partial t} = \nabla \times (\boldsymbol{v} \times \boldsymbol{B}_0)
\end{aligned} \tag{2.9}$$

In the vacuum region we have to solve

$$\nabla \times \delta \boldsymbol{B}_{\mathrm{v}} = 0 \quad \text{in} \quad \Omega_{\mathrm{v}} \ . \tag{2.10}$$

The auxiliary conditions that $\nabla \cdot \delta \boldsymbol{B} = 0$ and $\nabla \cdot \delta \boldsymbol{B}_{\mathrm{v}} = 0$ lead to the boundary conditions

$$\boldsymbol{n} \cdot \delta \boldsymbol{B} = \boldsymbol{n} \cdot \delta \boldsymbol{B}_{\mathrm{v}} \quad \text{at} \quad \Gamma_{\mathrm{p}} \tag{2.11}$$

and

$$\boldsymbol{n} \cdot \delta \boldsymbol{B}_{\mathrm{v}} = 0 \quad \text{at} \quad \Gamma_{\mathrm{v}} \ . \tag{2.12}$$

In addition, we have to impose the pressure balance condition at the plasma-vacuum interface, i.e.,

$$p + \boldsymbol{B}^2/2 = \boldsymbol{B}_{\mathrm{v}}^2/2 \quad \text{at} \quad \Gamma_{\mathrm{p}} \tag{2.13}$$

and that the tangential component of the electric field in the moving system be continuous (see *Kadomtsev* 1966 or *Appert* et al. 1982), which is written

$$\boldsymbol{n} \cdot \frac{\partial \delta \boldsymbol{B}_{\mathrm{v}}}{\partial t} = (\boldsymbol{B}_0 \cdot \nabla)(\boldsymbol{n} \cdot \boldsymbol{v}) \quad \text{at} \quad \Gamma_{\mathrm{p}} \ . \tag{2.14}$$

Let us introduce the Lagrangian displacement $\boldsymbol{\xi}(\boldsymbol{r}_0, t)$

$$\boldsymbol{r}(t) = \boldsymbol{r}_0 + \boldsymbol{\xi}(\boldsymbol{r}_0, t) \tag{2.15}$$

and

$$\boldsymbol{v}(\boldsymbol{r}_0, t) = \frac{\partial \boldsymbol{\xi}(\boldsymbol{r}_0, t)}{\partial t} \ , \tag{2.16}$$

where $\boldsymbol{r}_0$ is the unperturbed location of the fluid element which is at $\boldsymbol{r}$ at time t. The vectors $\boldsymbol{r}_0$ and $\boldsymbol{r}$ only differ by the infinitesimal quantity $\boldsymbol{\xi}$ so that we can replace $\boldsymbol{r}$ by $\boldsymbol{r}_0$ whenever $\boldsymbol{r}$ appears in coefficients of (2.9). These equations become, after a partial time integration,

$$\varrho_0 \frac{\partial^2 \xi(r,t)}{\partial t^2} = -\nabla \delta p + (\nabla \times \delta \boldsymbol{B}) \times \boldsymbol{B}_0 + \boldsymbol{J}_0 \times \delta \boldsymbol{B}$$

$$\delta p = -\xi \cdot \nabla p_0 - \gamma p_0 \nabla \cdot \xi \tag{2.17}$$

$$\delta \boldsymbol{B} = \nabla \times (\xi \times \boldsymbol{B}_0) \ .$$

Equation (2.10) remains unchanged. The boundary condition (2.14) now is written

$$\boldsymbol{n} \cdot \delta \boldsymbol{B}_{\mathrm{v}} = (\boldsymbol{B}_0 \cdot \nabla)(\boldsymbol{n} \cdot \xi) \quad \text{at} \quad \Gamma_{\mathrm{p}} \ . \tag{2.18}$$

By substituting the last two equations into the first one, we obtain the equation for the evolution of ξ $(\boldsymbol{r}, t)$

$$\varrho_0 \frac{\partial^2 \xi}{\partial t^2} = \mathscr{R}[\xi] \quad \text{in} \quad \Omega_{\mathrm{p}} \ , \tag{2.19}$$

where

$$\mathscr{R}[\xi] = \nabla(\gamma p_0 \nabla \cdot \xi + \xi \cdot \nabla p_0)$$
$$+ \boldsymbol{J}_0 \times [\nabla \times (\xi \times \boldsymbol{B}_0)] + \{\nabla \times [\nabla \times (\xi \times \boldsymbol{B}_0)]\} \times \boldsymbol{B}_0 \tag{2.20}$$

and $\mathscr{R}$ is a linear time-independent operator.

Equations (2.10, 19) and the boundary conditions admit solutions of the form

$$\xi(\boldsymbol{r}, t) = \xi(\boldsymbol{r})\, \mathrm{e}^{\mathrm{i}\omega t} \ , \tag{2.21}$$

where $\xi(\boldsymbol{r})$ can be chosen to be real. The equation of motion (2.19)

$$-\omega^2 \varrho_0 \xi(\boldsymbol{r}) = \mathscr{R}[\xi(\boldsymbol{r})] \ , \tag{2.22}$$

together with (2.10) and the boundary conditions constitute an eigenvalue problem.

2.4 Variational Formulation

This eigenvalue problem can be reformulated in variational form (*Lundquist* 1951; *Hain* et al. 1957; *Bernstein* et al. 1958)

$$\delta \mathscr{L} \equiv \delta[W_{\mathrm{p}} + W_{\mathrm{v}} - \omega^2 K] = 0 \ , \tag{2.23}$$

where

$$W_p = \frac{1}{2} \int_{\Omega_p} d^3r \{[\nabla \times (\xi \times B_0) + (\xi \cdot n)(J_0 \times n)]^2 - 2(\xi \cdot n)^2 (J_0 \times n) \cdot (B_0 \cdot n) n + \gamma p_0 (\nabla \cdot \xi)^2\} \qquad (2.24)$$

$$W_v = \frac{1}{2} \int_{\Omega_v} d^3r \delta B_v^2 \ , \qquad (2.25)$$

$$K = \frac{1}{2} \int_{\Omega_p} d^3r \varrho_0(r) \xi^2 \qquad (2.26)$$

and n is the vector normal to constant pressure surfaces.

In this formulation the pressure balance equation (2.13) constitutes a natural boundary condition. The remaining conditions then are (2.11), (2.12), and (2.18). In the particular case of a conducting wall put straight on the plasma surface, the fixed boundary condition is

$$n \cdot \xi = 0 \quad \text{at} \quad \Gamma_v = \Gamma_p \ . \qquad (2.27)$$

The plasma energy W_p and the kinetic energy $\omega^2 K$ are explicitly quadratic and symmetric forms in ξ. The vacuum energy W_v is a quadratic and symmetric form in δB_v, and is therefore implicitly quadratic and symmetric in ξ as well. As a consequence, the eigenvalue ω^2 is always real. In W_p the components of ξ tangential to the magnetic surfaces appear with only tangential derivatives. This can be seen by decomposing

$$\xi = (\xi \cdot n) n + \frac{\xi \cdot (n \times B_0)}{B_0^2} (n \times B_0) + \frac{\xi \cdot B_0}{B_0^2} B_0 \ . \qquad (2.28)$$

Since B_0 has no component in normal direction, i.e.,

$$n \cdot B_0 = 0 \qquad (2.29)$$

then,

$$\xi \times B_0 = (\xi \cdot n) B_0 + [\xi \cdot (n \times B_0)] n \ . \qquad (2.30)$$

It is now easy to show that no normal derivatives on the second and third components in (2.28) appear in (2.24). This means that these components may be discontinuous across a magnetic surface. In the variational form given in (2.23–26), the normal component $n \cdot \xi$ must be continuous and differentiable in any direction. The other two components of ξ must be continuous and differentiable only along magnetic surfaces.

Note, that the third component in (2.28) only appears in the $\nabla \cdot \xi$ term of (2.24).

2.5 Stability Considerations

In the variational form (2.23) the integral K, (2.26), is a quadratic and positive term. It plays the role of a normalization for ω^2 and does not influence the stability index τ which is characterized by the sign of ω^2:

$$\begin{aligned} &\omega^2>0: \tau=+1 \quad \text{stable} \\ &\omega^2=0: \tau=0 \quad \text{marginal} \\ &\omega^2<0: \tau=-1 \quad \text{unstable}\ , \end{aligned} \tag{2.31}$$

which is equivalent to

$$\begin{aligned} &W_p+W_v>0: \tau=+1 \quad \text{stable} \\ &W_p+W_v=0: \tau=0 \quad \text{marginal} \\ &W_p+W_v<0: \tau=-1 \quad \text{unstable}\ . \end{aligned} \tag{2.32}$$

2.6 Mechanical Analogon

In Fig. 2.2 we represent a mechanical analogon to understand the equilibrium and stability of a confined plasma. Also we try to show the limits of the ideal linear MHD model. Balls (A to G) are positioned at different places in an alpine landscape. Balls B to F are in equilibrium, ball A is not. Balls E and F are in a stable equilibrium state. Any infinitely small perturbation away from the equilibrium position will engender an oscillatory movement around positions x_3 for E and x_4 for F. The potential energies of the balls E and F increase when they are displaced slightly. Balls C and D are in unstable equilibrium states.

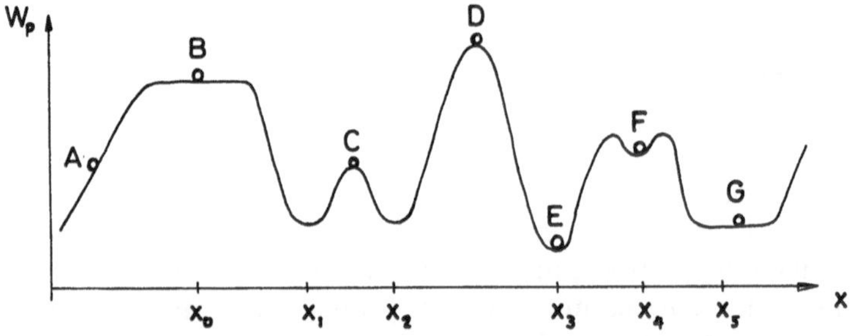

Fig. 2.2. Mechanical analogon to equilibrium and stability of a plasma. Positions x_1, x_2, x_3 and x_4 are stable equilibrium states. Positions x_0 and x_5 are marginally stable situations. Ball A is not in equilibrium, balls C and D are in an unstable equilibrium position

Any infinitely small perturbation will cause C to fall down towards x_1 or x_2, and D towards x_2 or x_3. The potential energies then decrease. Balls B and G do not change their potential energies when the equilibrium states are perturbed. The perturbation energies are entirely transformed into kinetic energies. They are in marginally stable positions.

These considerations are valid in the framework of ideal linear MHD theory. What happens when the equilibrium evolves and non-linear effects and non-ideal effects are included in the model? Marginal situations such as situation B can become unstable due to deformations of the plane by non-ideal effects such as resistivity. A small push out of the equilibrium position can bring ball F into an unstable situation and let it fall down towards x_3 or x_5. Such a locally stable equilibrium is referred to as a stability window. On the other hand, the unstable situation of ball C can, by a small change in equilibrium, eventually become stable (a so-called neighboring equilibrium) at positions x_1 or x_2.

3. Cylindrical Geometry

3.1 MHD Equations in Cylindrical Geometry

3.1.1 The AGV and Hain-Lüst Equations

In the case of a straight, circular, infinitely long plasma column (Fig. 3.1), the equilibrium quantities[1] only vary in the radial direction r. The equilibrium equations (2.5) can then be written as[2],

$$\frac{dp}{dr}+\frac{B_\theta}{r}\frac{d(rB_\theta)}{dr}+B_z\frac{dB_z}{dr}=0 \ , \tag{3.1}$$

where $p=p(r)$ denotes the plasma pressure and $\boldsymbol{B}=\boldsymbol{B}(r)=(0, B_\theta(r), B_z(r))$ is the magnetic field. Since the equilibrium solution is homogeneous in the poloidal (θ) and longitudinal (z) directions, the unknowns of the stability problem (2.17) can be Fourier-analysed in θ and z; a normal mode analysis of the type

$$\begin{pmatrix} \xi(r,\theta,z,t) \\ \delta p(r,\theta,z,t) \\ \delta\boldsymbol{B}(r,\theta,z,t) \end{pmatrix} = \begin{pmatrix} \xi(r) \\ \delta p(r) \\ \delta\boldsymbol{B}(r) \end{pmatrix} \exp(\mathrm{i}m\theta+\mathrm{i}kz+\mathrm{i}\omega t) \tag{3.2}$$

can be performed. After normal mode analysis the variables depend only on r, which varies from 0 to r_p in the plasma region Ω_p, and from r_p to r_v in the vacuum Ω_v. One can show that in the stability problem (2.17), the solutions for different wave numbers m and k decouple. This means that we may fix m and k in

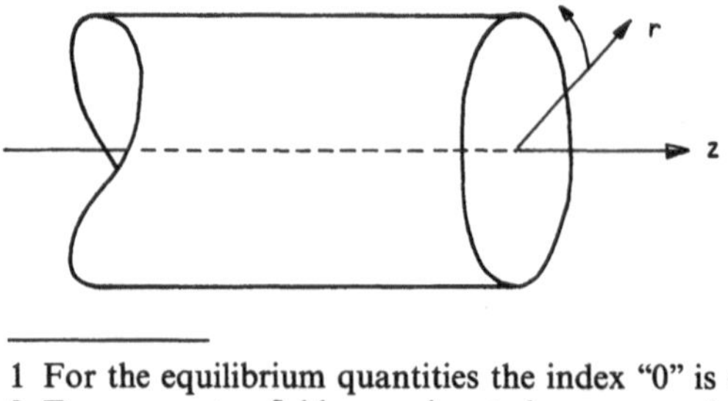

Fig. 3.1. The straight, circular, infinitely long plasma column

1 For the equilibrium quantities the index "0" is omitted
2 For any vector field $\boldsymbol{\eta}$ we denote by η_r, η_θ, and η_z its components in radial, poloidal and longitudinal direction

this chapter. Introducing the new variable δP (the perturbed total pressure)

$$\delta P = \delta p + \boldsymbol{B} \cdot \delta \boldsymbol{B} = \delta p + B_\theta \delta B_\theta + B_z \delta B_z \ , \tag{3.3}$$

the seven equations (2.17) become,

$$\begin{aligned}
&-\varrho\omega^2\xi_r - \mathrm{i}F\delta B_r + \delta P' + \frac{2B_\theta}{r}\delta B_\theta = 0 \\
&-\varrho\omega^2\xi_\theta - \mathrm{i}F\delta B_\theta + \frac{\mathrm{i}m}{r}\delta P - \frac{(rB_\theta)'}{r}\delta B_r = 0 \\
&-\varrho\omega^2\xi_z - \mathrm{i}F\delta B_z + \mathrm{i}k\delta P - B_z'\delta B_r = 0 \\
&\delta P - \mathrm{i}F(\xi_\theta B_\theta + \xi_z B_z) - \frac{2B_\theta^2}{r}\xi_r + (B_\theta^2 + B_z^2 + \gamma p)\nabla \cdot \boldsymbol{\xi} = 0 \\
&\delta B_r - \mathrm{i}F\xi_r = 0 \\
&\delta B_\theta - \mathrm{i}F\xi_\theta + r\left(\frac{B_\theta}{r}\right)'\xi_r + B_\theta \nabla \cdot \boldsymbol{\xi} = 0 \\
&\delta B_z - \mathrm{i}F\xi_z + B_z'\xi_r + B_z \nabla \cdot \boldsymbol{\xi} = 0 \ .
\end{aligned} \tag{3.4}$$

Here, $\varrho = \varrho(r) > 0$ is the mass density which can be chosen arbitrarily (the notation $'$ is the derivative with respect to r),

$$F = \boldsymbol{k} \cdot \boldsymbol{B} = \frac{m}{r} B_\theta + k B_z \tag{3.5}$$

and

$$\nabla \cdot \boldsymbol{\xi} = \frac{1}{r}(r\xi_r)' + \frac{\mathrm{i}m}{r}\xi_\theta + \mathrm{i}k\xi_z \ . \tag{3.6}$$

Note that five of the seven equations (3.4) are purely algebraic for the variables ξ_θ, ξ_z, δB_r, δB_θ, and δB_z. They can be eliminated without performing any differentiation. We obtain the so-called AGV equations which consist of two coupled first-order differential equations (*Appert* et al. 1974b) in the interval $(0, r_\mathrm{p})$

$$\begin{aligned}
&AS(r\xi_r)' = C_1 r\xi_r - rC_2\delta P \\
&AS(\delta P)' = \frac{1}{r} C_3 r\xi_r - C_1 \delta P \ .
\end{aligned} \tag{3.7}$$

The coefficients are

$$\begin{aligned}
&A=A(\omega,r)=\varrho\omega^2-F^2\\
&S=S(\omega,r)=\varrho\omega^2(B_\theta^2+B_z^2+\gamma p)-\gamma pF^2\\
&C_1=\frac{2B_\theta}{r}\left(\varrho^2\omega^4B_\theta-\frac{m}{r}FS\right)\\
&C_2=\varrho^2\omega^4-\left(k^2+\frac{m^2}{r^2}\right)S\\
&C_3=AS\left(A+2B_\theta\left(\frac{B_\theta}{r}\right)'\right)+\varrho\omega^2A\left(\frac{2B_\theta^2}{r}\right)^2-(S-\varrho\omega^2B_\theta^2)\left(\frac{2B_\theta F}{r}\right)^2 .
\end{aligned}\tag{3.8}$$

Eliminating δP by differentiating the first equation leads to the so-called Hain-Lüst equation (*Hain* and *Lüst* 1958; *Goedbloed* and *Hagebeuk* 1972).

$$\frac{d}{dr}\left[\frac{AS}{C_2r}\frac{d}{dr}(r\xi_r)\right]+\left[\frac{C_2C_3-C_1^2}{ASC_1}-r\frac{d}{dr}\left(\frac{C_1}{rC_2}\right)\right]\xi_r=0 .\tag{3.9}$$

At the magnetical axis we impose as regularity condition that

$$r\xi_r=0 \quad \text{for} \quad r=0 .\tag{3.10}$$

For the special case of a fixed boundary plasma, i.e., the wall is put on the plasma surface $r_v=r_p$, the boundary condition simply becomes

$$\xi_r=0 \quad \text{for} \quad r=r_p=r_v .\tag{3.11}$$

If there is a vacuum region limited by a conducting wall at $r_v>r_p$ we can fulfil (2.10) by the ansatz $[\Phi=\Phi(r)]$

$$\delta \boldsymbol{B}_v=\nabla\Phi ,\tag{3.12}$$

leading, in cylindrical geometry, to the Bessel equation

$$\frac{d^2\Phi}{dr^2}+\frac{2}{r}\frac{d\Phi}{dr}+\left(k^2+\frac{m^2}{r^2}\right)\Phi=0 .\tag{3.13}$$

At the conducting wall

$$\frac{d\Phi}{dr}(r_v)=0 .\tag{3.14}$$

At the plasma-vacuum interface Γ_p, (2.13) and (2.18) can be written

$$\begin{aligned}
&\delta P(r_p)=\mathrm{i}F(r_p)\Phi(r_p)\\
&\frac{d\Phi}{dr}(r_p)=\mathrm{i}F(r_p)\xi_r(r_p) .
\end{aligned}\tag{3.15}$$

Dividing these two equations leads to

$$F^2(r_p)\xi_r(r_p)=\frac{d\Phi}{dr}(r_p)\delta P(r_p)/\Phi(r_p) \ . \tag{3.16}$$

One possible way to solve (3.7, 13) numerically, subject to the conditions (3.10), (3.14–16) is to use a shooting method: For a guessed ω^2, one performs a Runge-Kutta integration from the magnetical axis to r_p imposing (3.10) and $\delta P(r=0)=\alpha$. In Ω_v another Runge-Kutta integration is performed from r_v to r_p, imposing (3.14) and $\Phi(r_v)=\beta$. In the matching condition (3.16) we can see that the two constants α and β for $\delta P(r=0)$ and for $\Phi(r_v)$ are arbitrary and can be chosen $\alpha=\beta=1$. Condition (3.16) is a condition for ω^2 which we have to adjust until (3.16) is satisfied.

Another method, presented in detail later in this chapter, is to solve the eigenvalue problem by a finite element approach.

3.1.2 Continuous Spectrum

Whenever $A=0$ or $S=0$, the system of (3.7) admits singular solutions (*Appert* et al. 1974b). More precisely, to all ω_0^2 for which there exists some $r=r_0$ in the interval $(0, r_p)$ such that

$$A=A(r_0,\omega_0^2)=\varrho(r_0)\omega_0^2-F^2(r_0)=0 \ , \tag{3.17}$$

or

$$S=S(r_0,\omega_0^2)=\varrho(r_0)\omega_0^2[B_\theta^2(r_0)+B_z^2(r_0)+\gamma p(r_0)]-\gamma p(r_0)F^2(r_0)=0 \ , \tag{3.18}$$

there exists a singular solution “localized” around $r=r_0$. The domain of these values ω_0^2 is the continuous spectrum of (3.7, 16). It consists of two, in general connected, real components. The component corresponding to smaller values of ω_0^2 is called “Slow mode continuum” ($S=0$) and the other is called “Alfvén Continuum” ($A=0$).

The Alfvén continuum ($A=0$) extends over

$$\underset{0\leq r\leq r_p}{\mathrm{Min}}\left(\frac{F^2}{\varrho}\right)\leq\omega_A^2\leq\underset{0\leq r\leq r_p}{\mathrm{Max}}\left(\frac{F^2}{\varrho}\right) , \tag{3.19}$$

and the Slow mode continuum ($S=0$) over

$$\underset{0\leq r\leq r_p}{\mathrm{Min}}\left(\frac{\gamma pF^2}{\varrho(B_\theta^2+B_z^2+\gamma p)}\right)\leq\omega_S^2\leq\underset{0\leq r\leq r_p}{\mathrm{Max}}\left(\frac{\gamma pF^2}{\varrho(B_\theta^2+B_z^2+\gamma p)}\right) . \tag{3.20}$$

These domains in ω_0^2 are always situated in the positive region of the ω^2 scale. For $F=0$ they reach the marginal point $\omega^2=0$. When the mass density ϱ drops to zero somewhere in the plasma (at the plasma surface for example), the continuous spectrum extends to infinity.

The "eigenmode" corresponding to an ω_0^2 lying in the continuous part of the spectrum shows a singular type of behavior. The Hain-Lüst equation (3.9) around the points $r=r_A$ or $r=r_S$, where $A(r_A)=0$ or $S(r_S)=0$, is of Fuchs's type when $r_A \neq 0$ and $r_S \neq 0$. It can be shown (*Barston* 1964) that, for finite ω, the radial component ξ_r develops a logarithmic singularity in $|r-r_A|$ or in $|r-r_S|$ around $r=r_A$ or $r=r_S$, whereas the tangential components ξ_θ and ξ_z exhibit $1/(r-r_A)$ or $1/(r-r_S)$ singularities.

At first glance, $C_2=0$ in the Hain-Lüst equation (3.9) also yields continua. However, this "singularity" is only apparent. It has been introduced by an elimination of the unknown δP. When $C_2=0$, the first equation in (3.7) decouples from the second one. The same is true when $C_3=0$. Then, the second equation decouples from the first one and, again, no new continua arise.

3.1.3 An Analytic Solution

The most simple straight circular equilibrium is a wall-constrained plasma without any current flowing along the cylinder. The boundary condition is given by (3.11). We assume constant mass density ϱ and a constant longitudinal magnetic field B_z. This equilibrium is characterized by

$$\begin{aligned} \boldsymbol{B}&=(0,0,1)\\ \boldsymbol{J}&=(0,0,0)\\ \varrho&=1\\ p&=\text{const}\ . \end{aligned} \tag{3.21}$$

The spectrum of this simple case can be calculated analytically (*Appert* et al. 1974a). The quantities A, S, C_1, C_2, and C_3 (3.8) in (3.7) become

$$\begin{aligned} A&=\omega^2-k^2\\ S&=\omega^2(1+\gamma p)-\gamma p k^2\\ C_1&=0\\ C_2&=\omega^4-\left(k^2+\frac{m^2}{r^2}\right)S\\ C_3&=A^2S\ . \end{aligned} \tag{3.22}$$

The Hain-Lüst equation is then

$$A\left\{\frac{d}{dr}\left[\frac{S}{rC_2}\frac{d}{dr}(r\xi_r)\right]+\xi_r\right\}=0\ . \tag{3.23}$$

There is an infinitely degenerate solution

$$A=0 \tag{3.24}$$

with the dispersion relation

$$\omega^2 = k^2 \ , \tag{3.25}$$

and

$$\begin{aligned} &\xi_z = \delta B_z = \delta P = \nabla \cdot \boldsymbol{\xi} = 0 \\ &\delta B_r = \mathrm{i} k \xi_r \\ &\delta B_\theta = \mathrm{i} k \xi_\theta \ . \end{aligned} \tag{3.26}$$

The incompressibility condition

$$\nabla \cdot \boldsymbol{\xi} = 0 \tag{3.27}$$

together with $\xi_z = 0$ relates ξ_r and ξ_θ by

$$D = (r\xi_r)' + \mathrm{i} m \xi_\theta = 0 \ . \tag{3.28}$$

This class of eigenmodes, fulfilling $A = 0$, is called the *Alfvén class*. In this case, the Alfvén continuum collapses to an infinitely degenerate eigenvalue.

There are two other classes of modes in the 1D ideal MHD spectrum with $A \neq 0$, $\xi_z \neq 0$, and $\nabla \cdot \boldsymbol{\xi} \neq 0$. They are given by the solution of the second-order ordinary differential equation

$$\frac{d}{dr}\left[\frac{S}{rC_2}\frac{d}{dr}(r\xi_r)\right] + \xi_r = 0 \ , \tag{3.29}$$

restricted by the condition (3.16). The solution is the radial derivative of a Bessel function J_m of order m:

$$\xi_r(r) = \mathrm{J}'_m(Kr) \ , \tag{3.30}$$

where

$$K^2 = A(\omega^2 - \gamma p k^2)/S \ . \tag{3.31}$$

The boundary condition (3.11) for a plasma of radius $r_\mathrm{p} = 1$

$$\xi_r(r = r_\mathrm{p} = 1) = \mathrm{J}'_m(K) = 0 \tag{3.32}$$

leads to the dispersion relation of the two other classes of eigenmodes:

$$\omega^2_{\substack{\mathrm{Fl} \\ \mathrm{Sl}}} = \tfrac{1}{2}(k^2 + \alpha^2_{ml})(1 + \gamma p)\left(1 \pm \sqrt{1 - \frac{4k^2\gamma p}{(k^2 + \alpha^2_{ml})(1 + \gamma p)}}\right) \ . \tag{3.33}$$

Here, α_{ml} denotes the lth zero of $d\mathrm{J}_m/dr$. The plus sign gives the dispersion relation for the so-called Fast modes (F) and the minus sign that for the Slow modes (S). One sees that ω_F^2 grows monotonically with l, i.e., with the number of zeros of the radial eigenfunction, ω_S^2 falls monotonically with increasing l towards the accumulation point

$$\omega_{S\infty}^2 = \gamma p k^2/(1+\gamma p) \ . \tag{3.34}$$

3.2 Six Test Cases

In this section we define six test cases which will be used to test and compare the different numerical methods presented in Chaps. 3–5. We will refer to them as test cases A to F.

3.2.1 Test Case A: Homogeneous Currentless Plasma Cylinder

Our first test case is the analytic solution of Sect. 3.1.3. In addition to the equilibrium parameters (3.21), we choose

$$p=0.05,\ r_p=1,\ \gamma=5/3,\ k=-0.5,\ m=1 \ . \tag{3.35}$$

For these parameters, the spectrum given by (3.25, 33) is shown in Fig. 3.2. One sees that the Alfvén class of modes is infinitely degenerate at $\omega_A^2 = k^2 = 0.25$ and the Fast modes show a Sturmian type of behavior up to $\omega^2 = \infty$. The Slow modes are very "dense". In order to make the Slow mode spectrum visible, the region around $\omega^2 = \omega_{S\infty}^2 = 1/52$ is magnified and shown again in Fig. 3.3. The exact distances from this accumulation point are listed in Table 3.1.

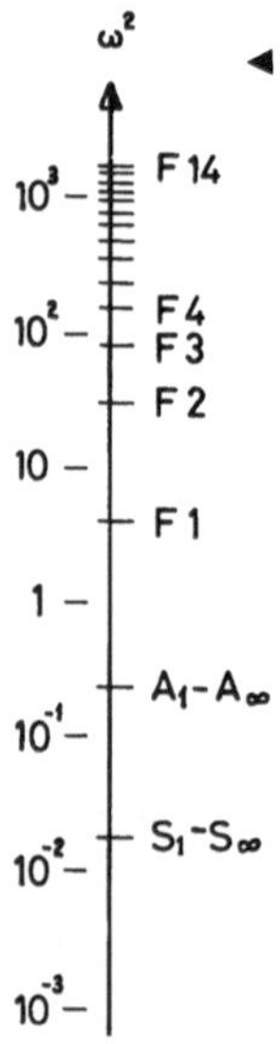

◄ Fig. 3.2

Fig. 3.2. Spectrum of the test case A in a logarithmic scale. F: Fast modes. A: Alfvén modes. S: Slow modes

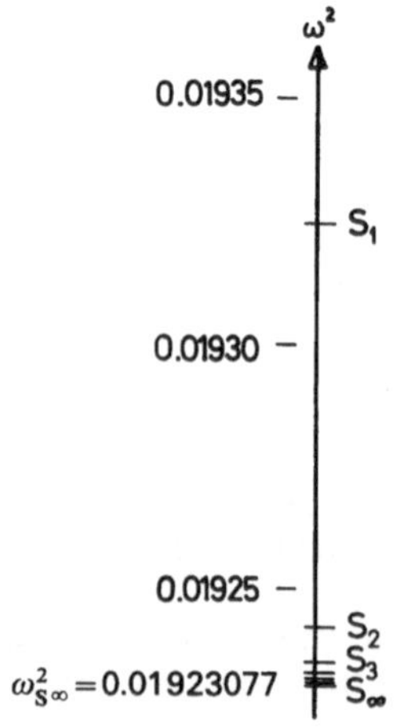

Fig. 3.3. Slow mode spectrum of the test case A in a linear scale

Table 3.1. Comparison of slow wave spectrum for the test case for $N=20$

Radial mode number n	$\omega^2-\omega_\infty^2$ analytic	$\omega^2-\omega_\infty^2$ polluted	$\omega^2-\omega_\infty^2$ unpolluted
1	$9{,}471\cdot10^{-5}$	$2{,}512\cdot10^{-4}$	$9{,}470\cdot10^{-5}$
2	$1{,}192\cdot10^{-5}$	$1{,}998\cdot10^{-4}$	$1{,}191\cdot10^{-5}$
3	$4{,}671\cdot10^{-6}$	$1{,}660\cdot10^{-4}$	$4{,}664\cdot10^{-6}$
4	$2{,}491\cdot10^{-6}$	$1{,}396\cdot10^{-4}$	$2{,}481\cdot10^{-6}$
5	$1{,}544\cdot10^{-6}$	$1{,}179\cdot10^{-4}$	$1{,}538\cdot10^{-6}$
6	$1{,}051\cdot10^{-6}$	$9{,}942\cdot10^{-5}$	$1{,}046\cdot10^{-6}$
7	$7{,}617\cdot10^{-7}$	*$9{,}484\cdot10^{-5}$	$7{,}576\cdot10^{-7}$
8	$5{,}773\cdot10^{-7}$	$8{,}354\cdot10^{-5}$	$5{,}747\cdot10^{-7}$
9	$4{,}527\cdot10^{-7}$	$6{,}975\cdot10^{-5}$	$4{,}517\cdot10^{-7}$
10	$3{,}644\cdot10^{-7}$	$5{,}771\cdot10^{-5}$	$3{,}655\cdot10^{-7}$
11	$2{,}997\cdot10^{-7}$	$4{,}720\cdot10^{-5}$	$3{,}033\cdot10^{-7}$
12	$2{,}508\cdot10^{-7}$	$3{,}806\cdot10^{-5}$	$2{,}573\cdot10^{-7}$
13	$2{,}129\cdot10^{-7}$	$3{,}015\cdot10^{-5}$	$2{,}227\cdot10^{-7}$
14	$1{,}831\cdot10^{-7}$	$2{,}345\cdot10^{-5}$	$1{,}966\cdot10^{-7}$
15	$1{,}591\cdot10^{-7}$	$1{,}846\cdot10^{-5}$	$1{,}770\cdot10^{-7}$
16	$1{,}395\cdot10^{-7}$	**$1{,}560\cdot10^{-5}$	$1{,}623\cdot10^{-7}$
17	$1{,}234\cdot10^{-7}$	$1{,}164\cdot10^{-5}$	$1{,}519\cdot10^{-7}$
18	$1{,}098\cdot10^{-7}$	$7{,}220\cdot10^{-6}$	$1{,}411\cdot10^{-7}$
19	$9{,}842\cdot10^{-8}$	$3{,}501\cdot10^{-6}$	$1{,}408\cdot10^{-7}$
20	$8{,}871\cdot10^{-8}$	$1{,}080\cdot10^{-6}$	–

* 1st mode
** 2nd mode (?)

3.2.2 Test Case B: Continuous Spectrum

Let us modify the equilibrium (3.21, 35) by replacing $\varrho=1$ by

$$\varrho(r)=1-\varepsilon r^2 \; . \tag{3.36}$$

A continuous spectrum results from such a mass density profile. The Alfvén continuum ($A=0$) extends over (3.19)

$$\begin{aligned} &k^2 \leqq \omega_A^2 \leqq k^2/(1-\varepsilon)\,, \qquad \varepsilon>0 \\ &k^2/(1+|\varepsilon|) \leqq \omega_A^2 \leqq k^2\,, \qquad \varepsilon<0 \end{aligned} \tag{3.37}$$

and the Slow mode continuum ($S=0$) over (3.20)

$$\begin{aligned} &\frac{\gamma p k^2}{1+\gamma p} \leqq \omega_A^2 \leqq \frac{\gamma p k^2}{(1+\gamma p)(1-\varepsilon)}\,, \qquad \varepsilon>0 \\ &\frac{\gamma p k^2}{(1+\gamma p)(1+|\varepsilon|)} \leqq \omega_A^2 \leqq \frac{\gamma p k^2}{1+\gamma p}\,, \qquad \varepsilon<0 \; . \end{aligned} \tag{3.38}$$

The effect of this mass density profile is sketched in Fig. 3.4.

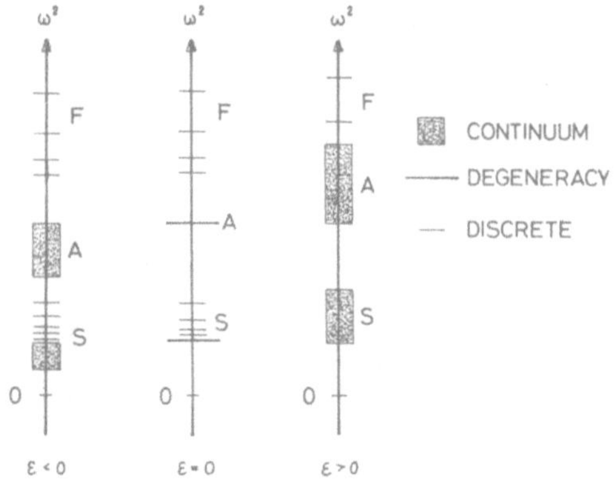

Fig. 3.4. Test case B: Effect of a density profile $\varrho(r)=1-\varepsilon r^2$ on test case A. For $\varepsilon \neq 0$ the degenerate points $\omega^2=\omega_A^2$ and $\omega^2=\omega_S^2$ open to continua. For $\varepsilon > 0$, modes associated with the discrete spectrum of the Slow modes are covered by the Slow mode continuum and the Alfvén continuum can overlap the Fast modes

3.2.3 Test Case C: Particular Free Boundary Mode

The equilibrium quantities for this test case are

$$\begin{aligned} &\boldsymbol{B}(r)=(0, B_\theta, 1) \\ &\boldsymbol{J}(r)=(0,0,j_z) \\ &\varrho(r)=1 \ . \end{aligned} \tag{3.39}$$

The plasma is surrounded by a vacuum region with a conducting wall at a distance of $r_v=R_v * r_p$, where $R_v \geq 1$ is a parameter to be chosen. For this test case $R_v=4$, where $B_\theta=rj_z/2$ with $j_z=$constant. We choose $\gamma=5/3$, $k=-0.5$, $m=1$, $r_p=1$ and the pressure p satisfies (3.1) with $p(1)=0$. The longitudinal current j_z is varied from 0 to 2. The spectrum as a function of j_z is shown in Fig. 3.5. For $j_z=0$ we have the free boundary spectrum of a homogeneous currentless plasma cylinder. The major difference between this and the fixed boundary case represented in Fig. 3.2 is in the lower eigenfrequencies of the Fast modes. As the current is turned on ($j_z>0$), the infinite degeneracy of the Alfvén modes opens to a discrete spectrum with an upper accumulation point at

$$\omega_{A\infty}^2=F^2/\varrho=\left(kB_z+\frac{m}{r}B_\theta\right)^2\Big/\varrho=\left(k+\frac{m}{2}j_z\right)^2=\frac{B_\theta^2}{r^2}(kq+m)^2 , \tag{3.40}$$

where $q\equiv rB_z/B_\theta$ is the safety factor. Around $\omega_{A\infty}^2=0$, i.e., around $j_z=1$, the Alfvén modes become unstable, i.e., $\omega_A^2<0$. Specifically, at $\omega_{A\infty}^2=0$, all Alfvén modes are unstable. As the current is further increased, all the Alfvén modes become stable again with the exception of the one called the "kink mode".

We remember that for $j_z=0$, the Slow mode spectrum is discrete with an accumulation point at the lower edge. As the current is turned on, this accumulation point opens to a continuous spectrum which eventually covers the discrete spectrum. For $j_z=1$, i.e., for $F=0$, the Slow mode spectrum degenerates to a point $\omega_S^2=0$.

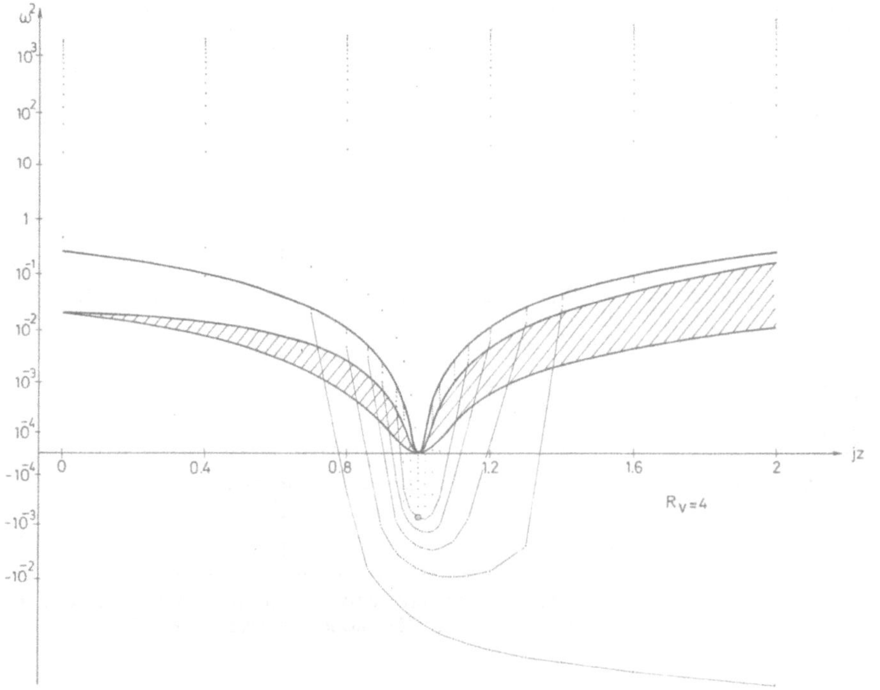

Fig. 3.5. Spectrum $\omega^2(j_z)$ for a free boundary case ($R_v=4$, $k=-0.5$, $m=1$). The dashed region is the Slow mode continuum. The thick upper line passing through $\omega^2=0$ at $j_z=1$ is the accumulation point of the Alfvén spectrum. There is an unstable region for $j_z>0.8$

As the current is increased, the eigenvalue of the lowest Fast mode decreases. This mode "penetrates" the Slow mode and the Alfvén spectrum around the point where $F=0$. From this point on, the lowest Alfvén mode changes its structure as the current is increased above $j_z=1$ and turns into the so-called "kink mode" which takes over the structure of the lowest Fast mode for $j_z<1$. This is a rigid displacement of the whole plasma column.

As our test case *C* we now choose a particular mode: the fifth unstable radial mode at $j_z=1$ (circle in Fig. 3.5) having an eigenvalue of $\omega^2=-9.018\times10^{-4}$. The global Bessel-function-like structure of this mode is recognizable in Fig. 3.6, where three components $\xi_r(r)$, $\xi_\theta(r)$, and $\xi_z(r)$ are represented. It is well suited to a study of the convergence behavior of different finite element approaches when applying them to global eigenmodes.

3.2.4 Test Case D: Unstable Region for $k=-0.2$, $m=1$

We consider a fixed boundary case with the same homogeneous profiles as those given in (3.39), leading to a similar spectrum as that given in Fig. 3.5. *Shafranov* (1970) showed that for such profiles, the distance $\omega^2-\omega^2_{A\infty}$ from the

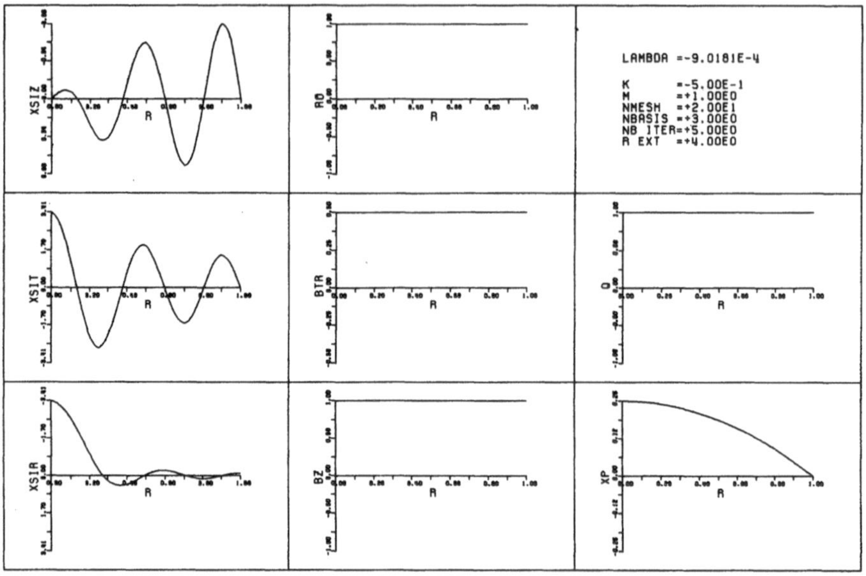

Fig. 3.6. Test case C: Profiles $\varrho(r)$, $B_\theta(r)/r$, $B_z(r)$, $g(r)$, and $p(r)$ and the radial structures $\xi_z(r)$, $\xi_\theta(r)$, and $\xi_r(r)$ of the fifth radial mode at $j_z=1$. The mode is marked in Fig. 3.5

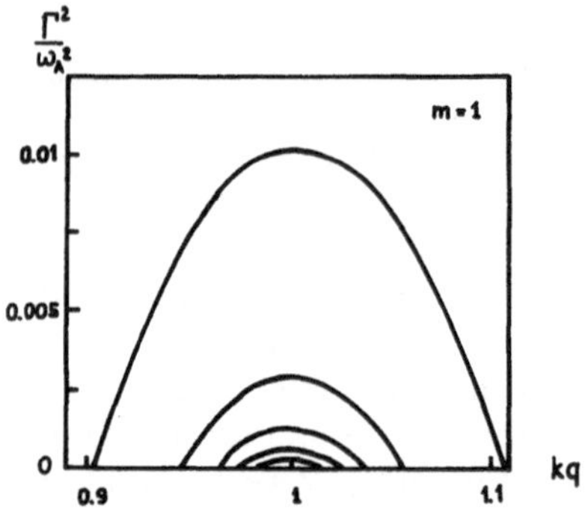

Fig. 3.7. Test case D: Normalized growth rate squared $\Gamma^2(kq)=-\omega^2$ of the five most unstable eigenmodes for $m=1$, $k=-0.2$, $B_z=1$, $\varrho=1$, $B_\theta/r=j_z/2$, $\omega_A^2=k^2=0.04$

accumulation point $\omega_{A\infty}^2=F^2/\varrho$ of the Alfvén modes is proportional to $k^2j_z^2/k_r^2$, where k_r is the radial wave number. In the unstable region, i.e., around $F=0$ or $j_z=-2k/m=0.4$ and $kq\equiv krB_z/B_\theta=1$, $\omega^2-\omega_{A\infty}^2\sim -k^4/m^2$. In order to construct a spectrum with a small unstable part, i.e., $\Gamma^2=-\omega^2>0$, where Γ^2 is the square of the growth rate, we have to choose small values of k and large values of m. In this test case D we choose $k=-0.2$ and $m=1$. The unstable part of the spectrum is shown in Fig. 3.7.

3.2.5 Test Case E: Unstable Region for $k=-0.2$, $m=2$

Compared to test case D, this case will be less unstable since $\Gamma^2\sim k^4/m^2$. The position, where $F=0$ corresponds now to $j_z=0.2$ and $kq=2$. The unstable part of this test case is shown in Fig. 3.8.

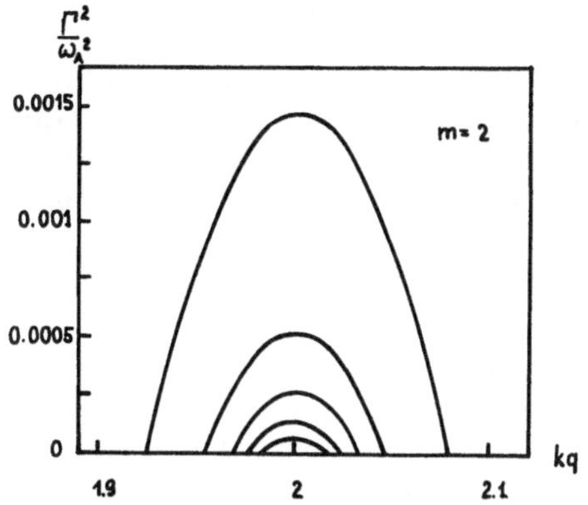

Fig. 3.8. Test case E: Same representation as Fig. 3.7 for $m=2$. Note, that Γ^2 has dropped by a factor of about 7

3.2.6 Test Case F: Internal Kink Mode

Finally, let us consider a more general equilibrium given by

$$\boldsymbol{B}(r)=\left(0, \frac{c_1 r}{1+c_2 r^2}, 1\right) , \tag{3.41}$$

leading to the pressure profile and the profile of the safety factor of

$$\begin{aligned} p(r) &= \frac{c_1^2}{c_2^2}\left[\frac{1}{(1+c_2^2 r^2)^2}-\frac{1}{(1+c_2^2)^2}\right] \\ kq(r) &= \frac{krB_z}{B_\theta}=\frac{1+c_2^2 r^2}{c_1}k \ . \end{aligned} \tag{3.42}$$

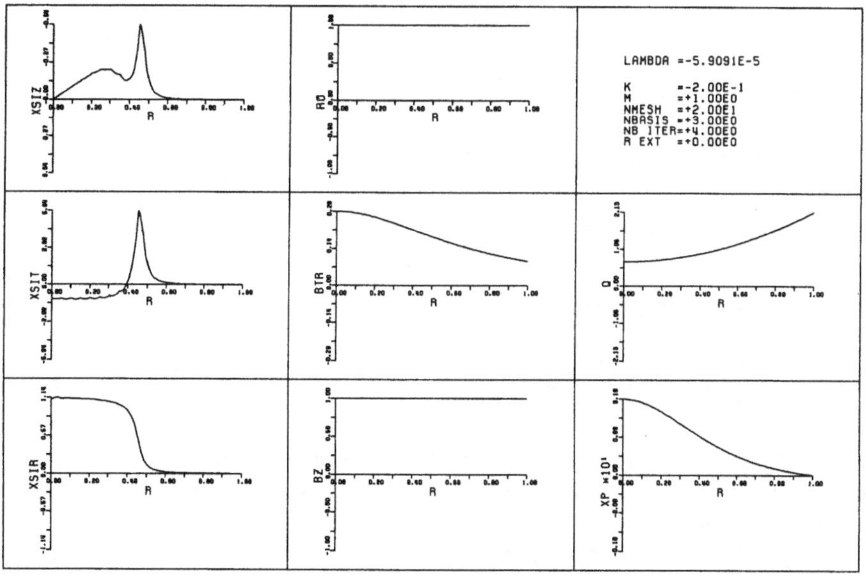

Fig. 3.9. Test case F: Profiles $\varrho(r)$, $B_\theta(r)/r$, $B_z(r)$, $q(r)$, and $p(r)$ and the radial structures $\xi_z(r)$, $\xi_\theta(r)$, and $\xi_r(r)$ of an unstable internal kink mode corresponding to the lowest eigenvalue

In our test case we choose $k=-0.2$, $c_1=2/7$, and $c_2=10/7$. The profiles of $B_z(r)$, $B_\theta(r)/r$, $\varrho(r)$, $kq(r)$, and $p(r)$ as well as the vector components ξ_r, ξ_θ, and ξ_z of the most unstable mode are plotted in Fig. 3.9. The square of the growth rate $\Gamma^2=5.9\times10^{-4}$ is very small. The eigenmode is localized and therefore quite different from that shown in Fig. 3.6.

3.3 Approximations

3.3.1 Conforming Finite Elements

Introducing the new variables

$$
\begin{aligned}
X &= r\xi_r \\
V &= \mathrm{i}(B_z\xi_\theta - B_\theta\xi_z)/B_z \\
Z &= \mathrm{i}r\xi_z/B_z \ ,
\end{aligned}
\tag{3.43}
$$

the plasma potential energy (2.24) may be written,

$$
\begin{aligned}
W_\mathrm{p} = \frac{1}{2}\int_0^{r_\mathrm{p}} \frac{dr}{r}\Bigg[& F^2X^2 + B_z^2(X'+mV)^2 + B_\theta^2\left(X' - kqV - \frac{2}{r}X\right)^2 \\
& + \gamma p(X'+mV+FZ)^2 - \frac{2B_\theta}{r^2}(rB_\theta)'X^2\Bigg] \ ,
\end{aligned}
\tag{3.44}
$$

where

$$
q = rB_z/B_\theta \tag{3.45}
$$

is the so-called safety factor. In the vacuum energy (2.25)

$$
W_\mathrm{v} = \frac{1}{2}\frac{\Phi(r_\mathrm{p})}{r_\mathrm{p}\Phi'(r_\mathrm{p})}F^2(r_\mathrm{p})X^2(r_\mathrm{p}) \ , \tag{3.46}
$$

the function $\Phi(r)$ is a solution of

$$
\frac{d^2\Phi}{dr^2} + \frac{2}{r}\frac{d\Phi}{dr} + \left(k^2 + \frac{m^2}{r^2}\right)\Phi = 0 \ , \tag{3.47}
$$

such that $\Phi'(r_\mathrm{v})=0$, where $'\equiv d/dr$ and r_v denotes the position of the conducting wall. The second condition to solve (3.47) is arbitrary since an arbitrary constant multiplying Φ does not alter expression (3.46).

Finally the kinetic energy (2.26) is

$$
W_\mathrm{K} = -\omega^2K = -\frac{\omega^2}{2}\int_0^{r_\mathrm{p}} \frac{\varrho\, dr}{r}[X^2 + (rV + B_\theta Z)^2 + B_z^2Z^2] \ . \tag{3.48}
$$

The variational formulation becomes

$$\delta\mathscr{L}=\delta(W_\mathrm{p}+W_\mathrm{v}-\omega^2K)=0 \ . \tag{3.49}$$

The regularity condition at the magnetic axis is

$$X=0 \quad \text{at} \quad r=0 \ . \tag{3.50}$$

Let us now introduce U to be the set of all functions $\boldsymbol{u}=(X, V, Z)\in(L^2(0, r_\mathrm{p}))^3$ with $dX/dr\in L^2(0, r_\mathrm{p})$ and $X(0)=0$. The variational formulation (3.44–49) is:

"Find real numbers ω^2 and non-trivial functions $\boldsymbol{u}=(X, V, Z)\in U$ such that $\delta\mathscr{L}=\delta(W_\mathrm{p}+W_\mathrm{v}-\omega^2K)=0$." (3.51)

As in Chap. 1 we build a finite-dimensional subspace U_h of U and solve the approximate problem:

"Find real numbers ω^2 and non-trivial functions $\boldsymbol{u}_h=(X_h, V_h, Z_h)\in U_h$ such that $\delta\mathscr{L}_h=\delta(W_\mathrm{p}+W_\mathrm{v}-\omega^2K)_h=0$." (3.52)

We have to note here that adding the vacuum energy W_v (3.46) is equivalent to imposing a mixed boundary condition

$$\alpha X(r_\mathrm{p})+X'(r_\mathrm{p})=0 \ . \tag{3.53}$$

In the case of an infinitely distant wall ($r_\mathrm{v}=\infty$), $W_\mathrm{v}=\alpha=0$ the remaining condition is a natural one. For a wall at the plasma surface, i.e., $r_\mathrm{v}=r_\mathrm{p}$, the vacuum contribution for a cylindrical plasma is $W_\mathrm{v}=\alpha=\infty$. The boundary condition for such a fixed boundary case is then,

$$X(r_\mathrm{p})=0 \ . \tag{3.54}$$

3.3.2 Non-Conforming Finite Elements

In addition to the variables X, V, and Z (3.43), we introduce a fourth one which is

$$\hat{X}=r\xi_r \ . \tag{3.55}$$

This new variable plays the same role as u_3 which we introduced in Sect. 1.2.4. The potential energies are written as

$$\begin{aligned}\hat{W}_\mathrm{p}=&\frac{1}{2}\int_0^{r_\mathrm{p}}\frac{dr}{r}\left[F^2X^2+B_z^2(\hat{X}'+mV)^2+B_\theta^2\left(\hat{X}'-kqV-\frac{2}{r}X\right)^2\right.\\&\left.+\gamma p(\hat{X}'+mV+FZ)^2-\frac{2B_\theta}{r^2}(rB_\theta)'X^2\right] ,\\ \hat{W}_\mathrm{v}=&\frac{1}{2}\frac{\Phi(r_\mathrm{p})}{r_\mathrm{p}\Phi'(r_\mathrm{p})}F^2(r_\mathrm{p})\hat{X}^2(r_\mathrm{p}) \ .\end{aligned} \tag{3.56}$$

Let us now define $\hat{U}$ to be the set of all functions $\hat{\boldsymbol{u}}=(X, \hat{X}, V, Z) \in (L^2(0, r_p))^4$, $\hat{X}' \equiv d\hat{X}/dr \in L^2(0, r_p)$ and $\hat{X}(0)=0$ such that

$$\int_0^{r_p} (X-\hat{X})\,\eta\, dr=0 \tag{3.57}$$

for all $\eta \in L^2(0, r_p)$. The variational formulation (3.49) is:

"Find real numbers ω^2 and non-trivial functions $\hat{\boldsymbol{u}}=(X, \hat{X}, V, Z) \in \hat{U}$ such that $\delta\hat{\mathscr{L}}=\delta(\hat{W}_p+\hat{W}_v-\omega^2 K)=0$." (3.58)

This new formulation (3.58) is strictly identical to that given in (3.52) since the integral (3.57) implies that $\hat{X}=X$.

We introduce a finite-dimensional space $\hat{U}_h$ in the same way as in Sect. 1.2.4. For all functions $\hat{\boldsymbol{u}}_h=(X_h, \hat{X}_h, V_h, Z_h) \in \hat{U}_h$ we have

$$\int_0^{r_p} (X_h-\hat{X}_h)\eta_h dr=0 \tag{3.59}$$

for all η_h in a finite-dimensional subspace of $L_2(0, r_p)$. The approximate problem is

"Find real numbers ω^2 and non-trivial functions $\hat{\boldsymbol{u}}_h=(X_h, \hat{X}_h, V_h, Z_h) \in \hat{U}_h$ such that $\delta\hat{\mathscr{L}}_h=\delta(\hat{W}_p+\hat{W}_v-\omega^2 K)_h=0$." (3.60)

3.4 Polluting Finite Elements

3.4.1 Hat Function Elements

The problem (3.52) is first solved by a finite element approach using hat functions as a basis for all the three components

$$\begin{pmatrix} X_h(r) \\ V_h(r) \\ Z_h(r) \end{pmatrix} = \sum_{j=0}^{N} \begin{pmatrix} X_j \\ V_j \\ Z_j \end{pmatrix} e_j(r) \ . \tag{3.61}$$

Here, X_j, V_j, and Z_j are the nodal values of the vector components at the nodal points r_j obtained by a discretization of the plasma region $0 \leqq r \leqq r_p$ into N, in general, non-equidistant intervals (Fig. 3.10). The basis functions $e_j(r)$ are shown in Fig. 1.2 and defined in (1.18, 19).

In the interval $r_j \leqq r \leqq r_{j+1}$, the vector components are represented by

$$\begin{pmatrix} X_h(r) \\ V_h(r) \\ Z_h(r) \end{pmatrix} = \begin{pmatrix} X_j \\ V_j \\ Z_j \end{pmatrix} \frac{r-r_{j+1}}{r_j-r_{j+1}} + \begin{pmatrix} X_{j+1} \\ V_{j+1} \\ Z_{j+1} \end{pmatrix} \frac{r-r_j}{r_{j+1}-r_j} \ . \tag{3.62}$$

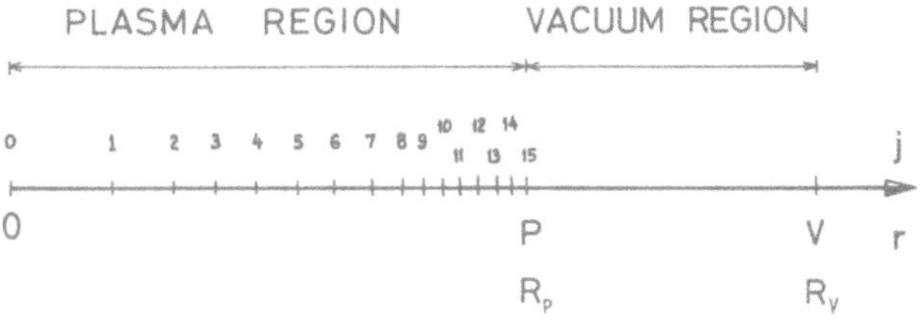

Fig. 3.10

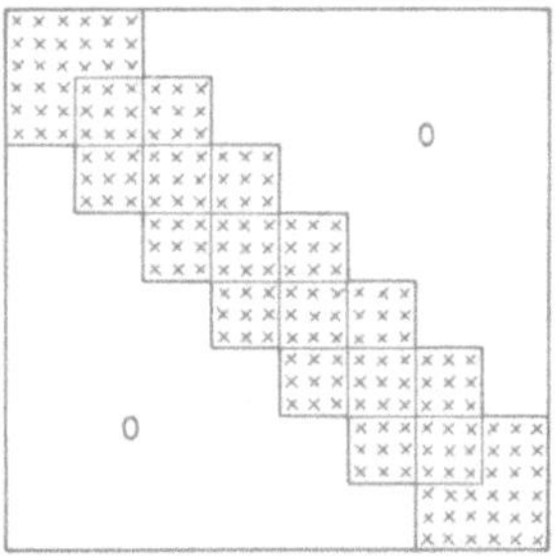

Fig. 3.11

Fig. 3.10. Example of a discretization into 15 non-equidistant intervals in the plasma region $0 \leq r \leq R_p$. The vacuum $R_p \leq r \leq R_v$ is integrated by a precise Runge-Kutta integration

Fig. 3.11. Form of matrices A and B for $N=7$ when linear finite elements $e(r)$ for all three vector components are taken. Each interval contributes a 6×6 submatrix

The variation over all unknowns X_j, V_j, and Z_j leads to a symmetric eigenvalue problem

$$Ax = \omega^2 Bx \ , \tag{3.63}$$

where B is positive definite and

$$x = (X_0, V_0, Z_0, \ldots, X_j, V_j, Z_j, \ldots, X_N, V_N, Z_N) \tag{3.64}$$

is the eigenvector. The matrix structures are those shown in Fig. 3.11. Multiplying (3.63) by x^T, the left-hand side $x^T Ax$ represents the discretized potential and vacuum energies, and the right-hand side $\omega^2 x^T Bx$ the discretized kinetic energy. In the eigenvalue problem (3.63), ω^2 denotes the eigenvalue. Positive ω^2 gives a stable oscillatory displacement, negative ω^2 leads to an unstable growing mode as discussed in Sect. 2.5. The case $\omega^2 = 0$ is referred to as marginally stable.

3.4.2 Application to Test Case A

Let us apply this hat function approach to the test case A (3.35). The results are shown in Fig. 3.12, where we plot all the eigenvalues ω^2 as a function of the number of radial intervals N. We always find N Slow, N Alfvén, and N Fast modes.

For $N=2$ to $N=20$, the Fast modes are easily recognized. The mode structures and the eigenvalues are very close to the analytical predictions.

The Alfvén class should be infinitely degenerate with an eigenvalue $\omega^2 = k^2$. What we find numerically is a spectrum similar to that of the Fast modes with eigenfunctions of Bessel type. The more intervals we take, the more the

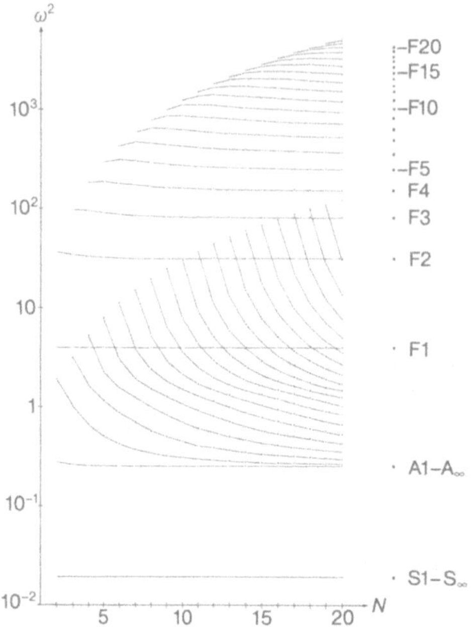

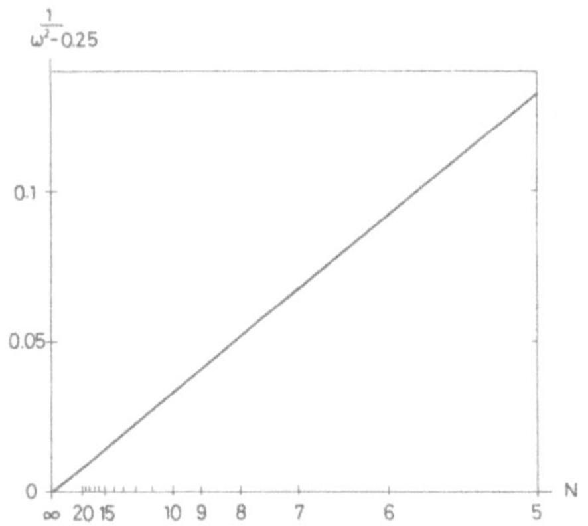

Fig. 3.12 **Fig. 3.13**

Fig. 3.12. Spectrum of test case A as a function of the number of intervals using linear finite elements for all three vector components. The analytic solution is given at the right hand side border. Instead of the infinitively degenerate solution $\omega^2 = k^2 = 0.25$, the Alfvén modes show a "Sturmian" behaviour, and interfere with the Fast modes. This phenomenon is called "spectral pollution"

Fig. 3.13. Inverse distance $\omega^2 - 0.25$ of the uppermost Alfvén mode as a function of the number of intervals N(scale $\sim 1/N^2$). The straight line towards 0 means that the Alfvén spectrum tends quadratically towards ∞ with increasing N

spectrum spreads. This phenomenon is studied in Fig. 3.13. Here, we represent the inverse of the distance between the uppermost highly oscillating Alfvén mode A_N and the exact value $\omega^2 = k^2$ as a function of $1/N^2$. What we see is that the "creation curve" of the most oscillating Alfvén modes A_N tends to infinity as $N \to \infty$. Instead of an infinitely degenerate solution $\omega^2 = k^2$ we obtain a "Sturmian" spectrum which extends to infinity. Each individual mode with a fixed radial mode number converges quadratically towards $\omega^2 = k^2 = 0.25$, but the eigenvalues of the newly created highly oscillating modes tend to infinity.

A similar behavior is seen for the Slow modes which are "very dense" around the accumulation point $\omega^2_{S\infty} = 1/52 = 0.019230769$. To make the effect visible we have magnified the region around $\omega^2_{S\infty}$ and represent it separately in Fig. 3.14. Again ω^2 is shown as a function of the number of radial intervals N. With increasing N the most oscillatory mode is more and more distant from the exact value which should be close to $\omega^2_{S\infty}$. This distance seems to be $\leqq 2.71 \times 10^{-4}$. The recognition of the modes is very difficult since, as N increases, each

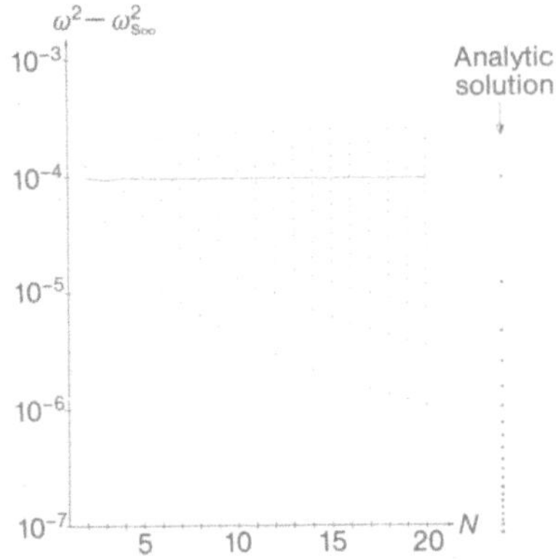

Fig. 3.14

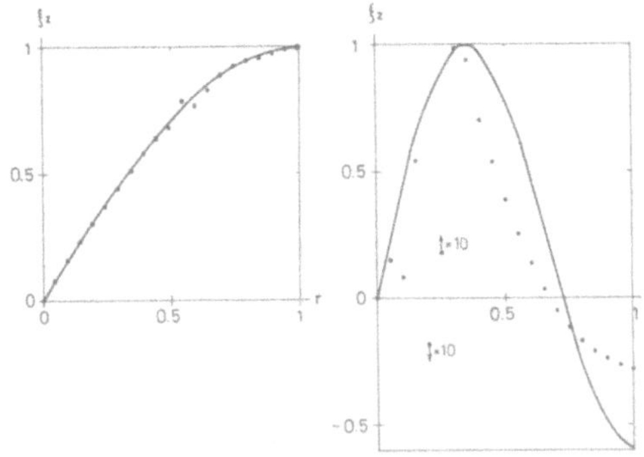

Fig. 3.15

Fig. 3.14. Spectral pollution of the Slow modes for test case A. Only the first radial mode with $\omega_S^2\infty = 0.019325$ can be detected with certainty

Fig. 3.15. Radial structures of ξ_z of the first ($S1$) and second ($S2$) Slow mode of the analytic (—) and numerical (···) solutions using linear elements for all three components. Pollution leads to strong interactions with higher radial modes which cross the spectrum

newly created eigenmode has to cross the whole previously created spectrum. It is possible to identify the first Slow mode S1. The second radial mode S2, however, is already very difficult to identify. The radial structures of the longitudinal component ξ_z of these two eigenmodes are shown in Fig. 3.15 for $N=20$. We see that S1 agrees well with the analytically calculated mode. It shows a small irregularity around $r=0.57$. The mode which we defined to be S2, already has a different shape from the analytic one. Strong interactions with higher radial modes lead to jumps in the eigenfunction for $r<0.3$. The slow modes should be Bessel functions but are solutions with jumps, whereas, in the Alfvén class, the eigenmodes are regular Bessel-like functions instead of jumping ones. In both classes of modes, the approximated spectrum does not converge towards the analytic one. This is the phenomenon that we called "spectrum pollution" in the first chapter.

3.5 Conforming Non-Polluting Finite Elements

3.5.1 Linear Elements

In the upper approach (linear finite elements for all three vector components), the spectrum pollution comes from the fact that the conditions (3.27, 28), which are

$$\begin{aligned} \frac{dX}{dr} + mV &= 0 \\ Z &= 0 \end{aligned} \tag{3.65}$$

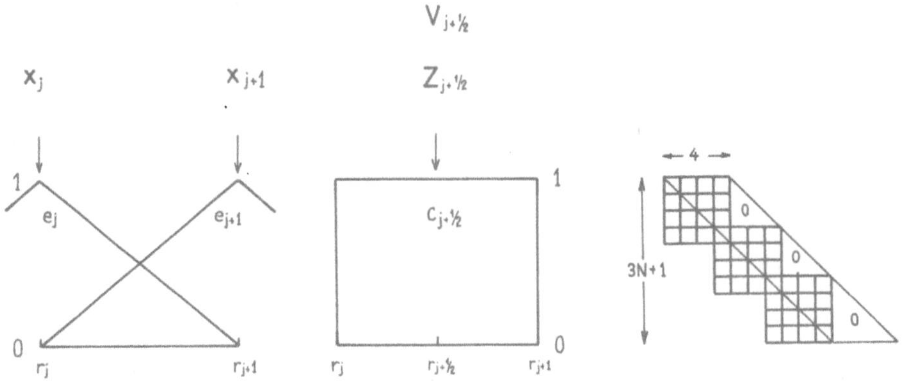

Fig. 3.16. The linear finite elements e_j and e_{j+1} for the normal component X_h, and the piecewise constant elements $c_{j+1/2}$ for the two tangential components V_h and Z_h. The symmetric band matrices A and B have a half band width of 4. The total number of unknowns is $3N+1$

and which determine the Alfvén spectrum of the test case A, cannot be satisfied everywhere. With linear elements for X_h and V_h, dX_h/dr becomes piecewise constant and the incompressibility condition of (3.65) can only be satisfied at one point of each interval. We subdivide the interval $[0, r_p]$ into N, in general, non-equidistant intervals (Fig. 3.10). To solve the problem (3.52) the least regular basis functions to represent the three components are (*Appert* et al. 1975a),

$$X_h(r) = \sum_{j=1}^{N} X_j e_j(r)$$
$$V_h(r) = \sum_{j=1}^{N} V_{j-1/2} c_{j-1/2}(r) \tag{3.66}$$
$$Z_h(r) = \sum_{j=1}^{N} Z_{j-1/2} c_{j-1/2}(r) \ .$$

The basis function $e_j(r)$ is a hat function defined in (1.18, 19) and shown in Figs. 1.2, 3.16. The basis function $c_{j+1/2}(r)$ is defined in (1.39) and shown in Fig. 3.16. In the interval $r_j \leqq r \leqq r_{j+1}$ the functional variations of X_h, V_h, and Z_h are

$$X_h(r) = X_j \frac{r-r_{j+1}}{r_j-r_{j+1}} + X_{j+1} \frac{r-r_j}{r_{j+1}-r_j}$$
$$V_h(r) = V_{j+1/2} \tag{3.67}$$
$$Z_h(r) = Z_{j+1/2} \ .$$

This means that X_h is piecewise linear and V_h and Z_h are piecewise constant in each interval. With such an approach, the incompressibility condition (3.65) can be fulfilled everywhere on each internal (r_j, r_{j+1}).

With these elements the band matrices A and B of the matrix eigenvalue problem (3.63) have the form given in Fig. 3.16. The half-band-width is 4 and the total number of unknowns is $3N+1$.

3.5.2 Quadratic Elements

Second-order Lagrange elements S_2^1 (1.71–72) for the discretization of X_h and piecewise linear discontinuous elements S_1^0 for V_h and Z_h (1.70) are also in a function class which fulfil (3.65). In one interval $r_j \leqq r \leqq r_{j+1}$ all the non-zero basis functions are shown in Fig. 3.17. The functional dependences of $X_h(r)$, $V_h(r)$, and $Z_h(r)$ in $r_j \leqq r \leqq r_{j+1}$ are

$$X_h(r) = X_j \frac{(r-r_{j+1/2})(r-r_{j+1})}{(r_j-r_{j+1/2})(r_j-r_{j+1})} + X_{j+1/2} \frac{(r-r_j)(r-r_{j+1})}{(r_{j+1/2}-r_j)(r_{j+1/2}-r_{j+1})}$$
$$+ X_{j+1} \frac{(r-r_j)(r-r_{j+1/2})}{(r_{j+1}-r_j)(r_{j+1}-r_{j+1/2})} \tag{3.68}$$
$$\begin{pmatrix} V_h(r) \\ Z_h(r) \end{pmatrix} = \begin{pmatrix} V_j^{\mathrm{R}} \\ Z_j^{\mathrm{R}} \end{pmatrix} \frac{r-r_{j+1}}{r_j-r_{j+1}} + \begin{pmatrix} V_{j+1}^{\mathrm{L}} \\ Z_{j+1}^{\mathrm{L}} \end{pmatrix} \frac{r-r_j}{r_{j+1}-r_j} .$$

We have to note here that at each mesh point there are two nodal values for V_h and Z_h, one which goes with the interval on the left-hand side (upper index L) and one which goes with the interval on the right-hand side (upper index R).

For this approach, the symmetric band matrices A and B are shown in the lower part of Fig. 3.17. They have a half-width of 7 and the total number of unknowns is $6N+1$.

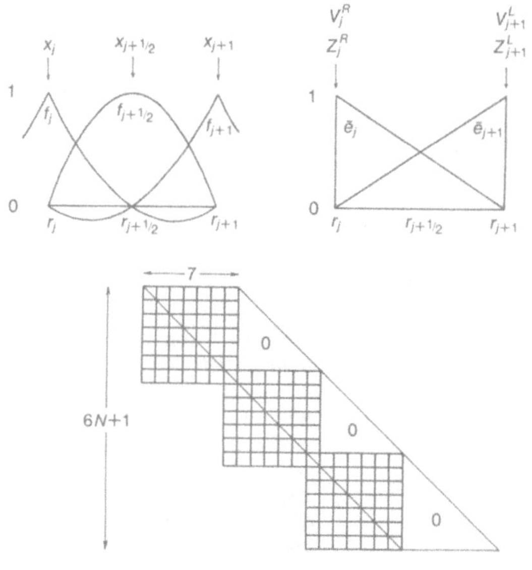

Fig. 3.17. Quadratic finite elements for $X_h(r)$, and discontinuous linear elements for $V_h(r)$ and $Z_h(r)$. Note that V_h and Z_h jump across the intervals. At $r=r_j$ the values of V_h and Z_h corresponding to the left hand side interval $[r_{j-1}, r_j]$ are V_j^{L} and Z_j^{L}. They are V_j^{R} and Z_j^{R} when they correspond to the right hand side interval $[r_j, r_{j+1}]$. For this approach, the symmetric band matrices A and B (*shown below*) have a half width of 7. The total number of unknowns is $6N+1$

3.5.3 Third-Order Lagrange Elements

An "ecological" (i.e., non-polluting) approach is guaranteed when one takes elements S_3^1 (1.71) to represent X_h and second-order discontinuous elements S_2^0 (1.70) for V_h and Z_h. In the interval $r_j \leqq r \leqq r_{j+1}$ all the non-zero basis functions are shown in Fig. 3.18. The functional dependences of X_h, V_h, and Z_h in $r_j \leqq r \leqq r_{j+1}$ are

$$\begin{aligned} X_h(r) = {} & X_j \frac{(r-r_{j+1/3})(r-r_{j+2/3})(r-r_{j+1})}{(r_j-r_{j+1/3})(r_j-r_{j+2/3})(r_j-r_{j+1})} \\ & + X_{j+1/3} \frac{(r-r_j)(r-r_{j+2/3})(r-r_{j+1})}{(r_{j+1/3}-r_j)(r_{j+1/3}-r_{j+2/3})(r_{j+1/3}-r_{j+1})} \\ & + X_{j+2/3} \frac{(r-r_j)(r-r_{j+1/3})(r-r_{j+1})}{(r_{j+2/3}-r_j)(r_{j+2/3}-r_{j+1/3})(r_{j+2/3}-r_{j+1})} \\ & + X_{j+1} \frac{(r-r_j)(r-r_{j+1/3})(r-r_{j+2/3})}{(r_{j+1}-r_j)(r_{j+1}-r_{j+1/3})(r_{j+1}-r_{j+2/3})} \,, \end{aligned} \tag{3.69}$$

$$\begin{aligned} \begin{pmatrix} V_h(r) \\ Z_h(r) \end{pmatrix} = {} & \begin{pmatrix} V_j^{\mathrm{R}} \\ Z_j^{\mathrm{R}} \end{pmatrix} \frac{(r-r_{j+1/2})(r-r_{j+1})}{(r_j-r_{j+1/2})(r_j-r_{j+1})} + \begin{pmatrix} V_{j+1/2} \\ Z_{j+1/2} \end{pmatrix} \frac{(r-r_j)(r-r_{j+1})}{(r_{j+1/2}-r_j)(r_{j+1/2}-r_{j+1})} \\ & + \begin{pmatrix} V_{j+1}^{\mathrm{L}} \\ Z_{j+1}^{\mathrm{L}} \end{pmatrix} \frac{(r-r_j)(r-r_{j+1/2})}{(r_{j+1}-r_j)(r_{j+1}-r_{j+1/2})} \,. \end{aligned}$$

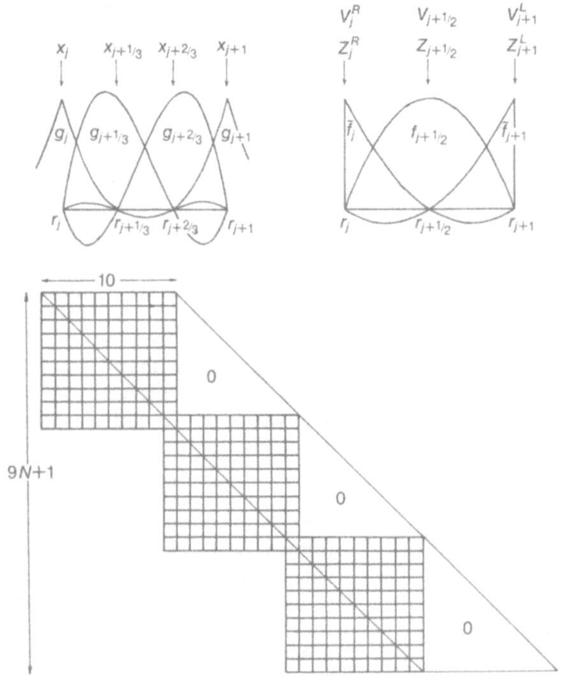

Fig. 3.18. Cubic Lagrange polynomials (*left*) for $X_h(r)$, quadratic (*right*) for $V_h(r)$ and $Z_h(r)$. Again V_h and Z_h jump across the intervals. For this approach the symmetric band matrices A and B shown below have a halfwidth of 10. The total number of unknowns is $9N+1$

Again we have to note that V_h and Z_h jump across the intervals with nodal values indexed with L for the left-hand side interval, and nodal values indexed with R for the right-hand side interval.

For cubic Lagrange elements, the symmetric band matrices A and B (Fig. 3.18) have a half-width of 10 and the total number of unknowns is $9N+1$.

3.5.4 Cubic Hermite Elements

S_3^2 (1.75, 76) for the discretization of X_h and second-order Lagrange elements S_2^1 (1.71) for V_h and Z_h guarantee an unpolluted approach. In the interval $r_j \leqq r \leqq r_{j+1}$ all of the non-zero basis functions are shown in Fig. 3.19. The functional dependences of X_h, V_h, and Z_h in $r_j \leqq r \leqq r_{j+1}$ are

$$\begin{aligned} X_h(r) = & -X_j \frac{(r-r_{j+1})^2(3r_j - r_{j+1} - 2r)}{(r_{j+1}-r_j)^3} + X_j^* \frac{(r-r_j)(r-r_{j+1})^2}{(r_{j+1}-r_j)^2} \\ & + X_{j+1} \frac{(r-r_j)^2(3r_{j+1}-r_j-2r)}{(r_{j+1}-r_j)^3} + X_{j+1}^* \frac{(r-r_j)^2(r-r_{j+1})}{(r_{j+1}-r_j)^2} \,, \end{aligned}$$

$$\begin{aligned} \begin{pmatrix} V_h(r) \\ Z_h(r) \end{pmatrix} = & \begin{pmatrix} V_j \\ Z_j \end{pmatrix} \frac{(r-r_{j+1/2})(r-r_{j+1})}{(r_j-r_{j+1/2})(r_j-r_{j+1})} \\ & + \begin{pmatrix} V_{j+1/2} \\ Z_{j+1/2} \end{pmatrix} \frac{(r-r_j)(r-r_{j+1})}{(r_{j+1/2}-r_j)(r_{j+1/2}-r_{j+1})} \\ & + \begin{pmatrix} V_{j+1} \\ Z_{j+1} \end{pmatrix} \frac{(r-r_j)(r-r_{j+1/2})}{(r_{j+1}-r_j)(r_{j+1}-r_{j+1/2})} \,. \end{aligned} \tag{3.70}$$

Here, X_j and X_{j+1} are the nodal values of X_h at r_j and r_{j+1}, whereas X_j^* and X_{j+1}^* are the nodal values of dX_h/dr at r_j and r_{j+1}.

For Hermite elements, the symmetric band matrices A and B (Fig. 3.19) have a half-band-width of 10 and the total number of unknowns is $6N+4$.

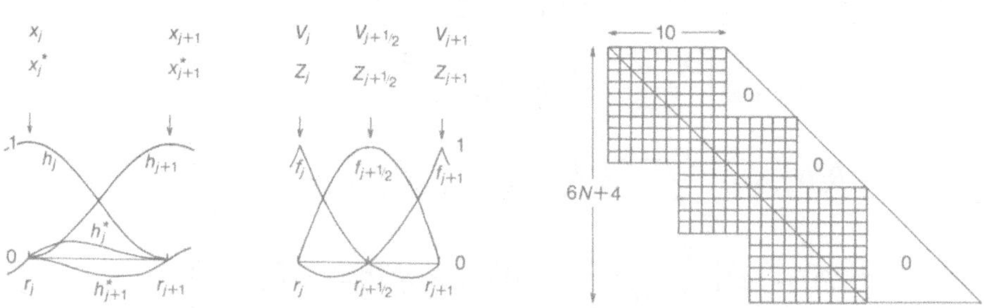

Fig. 3.19. Cubic Hermite polynomials for $X_h(r)$ and quadratic elements for $V_h(r)$ and $Z_h(r)$ in the internal $[r_j, r_{j+1}]$. For this approach, the symmetric band matrices A and B have a halfwidth of 10. The total number of unknowns is $6N+4$

3.6 Non-Conforming Non-Polluting Elements

3.6.1 Linear Elements

Let us now solve Problem (3.60), where we added a new variable $\hat{X}_h$ (3.55) introduced to represent $\hat{X}$ when it appears with the radial derivative $d\hat{X}/dr$ in the variational formulation. This new variable is related to X_h through condition (3.59). We first consider the *lowest order elements*:

$$\hat{X}_h(r) = \sum_{j=1}^{N} X_j e_j(r), \quad \begin{pmatrix} X_h(r) \\ V_h(r) \\ Z_h(r) \end{pmatrix} = \sum_{j=1}^{N} \begin{pmatrix} X_{j-1/2} \\ V_{j-1/2} \\ Z_{j-1/2} \end{pmatrix} c_{j-1/2}(r) \ . \tag{3.71}$$

The linear basis functions $e_j(r)$ are defined in (1.18, 19) and shown in Figs. 1.2, 3.16. The piecewise constant basis functions $c_{j-1/2}(r)$ are defined in (1.39) and shown in Fig. 3.16. In condition (3.59) we choose the piecewise constant basis functions $c_{j-1/2}$ as test functions η_h. The condition then becomes for all $l = 1, \ldots, N$

$$\int_0^{r_p} \sum_{j=1}^{N} [X_{j-1/2} c_{j-1/2}(r) - X_j e_j(r)] c_{l-1/2}(r) dr = 0 \ . \tag{3.72}$$

It is possible to eliminate the variables at the half-interval points. In the interval $[r_j, r_{j+1}]$ this variable is

$$X_{j+1/2} = \frac{X_j + X_{j+1}}{2} \ . \tag{3.73}$$

The radial derivative $d\hat{X}_h/dr$ is also very simple. In $[r_j, r_{j+1}]$ it is

$$\frac{d\hat{X}_h}{dr} = \frac{X_{j+1} - X_j}{r_{j+1} - r_j} \ . \tag{3.74}$$

The energies in (3.58) are then

$$\begin{aligned} \hat{W}_p = \frac{1}{2} \sum_{j=1}^{N} \frac{r_j - r_{j-1}}{r_{j-1/2}} \Bigg\{ & F^2(r_{j-1/2}) \left[\frac{X_{j-1} + X_j}{2} \right]^2 \\ & + B_z^2(r_{j-1/2}) \left[\frac{X_j - X_{j-1}}{r_j - r_{j-1}} + m V_{j-1/2} \right]^2 \\ & + B_\theta^2(r_{j-1/2}) \left[\frac{X_j - X_{j-1}}{r_j - r_{j-1}} - kq(r_{j-1/2}) V_{j-1/2} - \frac{X_{j-1} + X_j}{r_{j-1/2}} \right]^2 \\ & + \gamma p(r_{j-1/2}) \left[\frac{X_j - X_{j-1}}{r_j - r_{j-1}} + m V_{j-1/2} + F(r_{j-1/2}) Z_{j-1/2} \right]^2 \\ & - \frac{2 B_\theta^2(r_{j-1/2})}{r_{j-1/2}^2} \left(\frac{r_{j-1/2} B_\theta'(r_{j-1/2})}{B_\theta(r_{j-1/2})} + 1 \right) \left[\frac{X_{j-1} + X_j}{2} \right]^2 \Bigg\} \end{aligned} \tag{3.75}$$

$$\hat{W}_v = \frac{1}{2}\frac{\Phi(r_p)}{r_p\Phi'(r_p)}F^2(r_p)X_N^2$$

$$W_K = -\frac{\omega^2}{2}\sum_{j=1}^{N}\frac{r_j - r_{j-1}}{r_{j-1/2}}\varrho(r_{j-1/2})\left\{\left[\frac{X_{j-1}+X_j}{2}\right]^2\right.$$

$$\left. + [r_{j-1/2}V_{j-1/2} + B_\theta(r_{j-1/2})Z_{j-1/2}]^2 + B_z^2(r_{j-1/2})Z_{j-1/2}^2\right\} .$$

The integral is replaced by a sum over the integrands at the mid points, times the interval size. We call this approach a "finite hybrid element" approach (*Gruber* 1978).

3.6.2 Quadratic Elements

Using non-conforming, non-polluting second-order Lagrange elements, the unknowns in the interval $[r_j, r_{j+1}]$ become (Fig. 3.17)

$$\hat{X}_h(r) = \frac{2(r-r_{j+1/2})(r-r_{j+1})}{(r_{j+1}-r_j)^2}X_j - \frac{4(r-r_j)(r-r_{j+1})}{(r_{j+1}-r_j)^2}X_{j+1/2}$$
$$+\frac{2(r-r_j)(r-r_{j+1/2})}{(r_{j+1}-r_j)^2}X_{j+1} \tag{3.76}$$
$$\begin{pmatrix} X_h(r) \\ V_h(r) \\ Z_h(r) \end{pmatrix} = \begin{pmatrix} X_j^R \\ V_j^R \\ Z_j^R \end{pmatrix}\frac{r-r_{j+1}}{r_j - r_{j+1}} + \begin{pmatrix} X_{j+1}^L \\ V_{j+1}^L \\ Z_{j+1}^L \end{pmatrix}\frac{r-r_j}{r_{j+1}-r_j} ,$$

where the notation is the same as for (3.68). It is possible to eliminate intervalwise the two unknowns X_j^R and X_{j+1}^L by making the first two moments of (3.59) be zero. This means that

$$\int_{r_j}^{r_{j+1}} [\hat{X}_h(r) - X_h(r)]\,dr = 0$$
$$\int_{r_j}^{r_{j+1}} [\hat{X}_h(r) - X_h(r)]\,r\,dr = 0 , \tag{3.77}$$

leading to

$$X_h(r) = \frac{1}{6}(X_j + 4X_{j+1/2} + X_{j+1}) + \frac{r - r_{j+1/2}}{r_{j+1}-r_j}(X_{j+1} - X_j) . \tag{3.78}$$

3.6.3 Lagrange Cubic Elements

Non-conforming, non-polluting third-order Lagrange elements are characterized by the following functional dependences of $X_h(r)$, $\hat{X}_h(r)$, $V_h(r)$, and $Z_h(r)$ in the interval $[r_j, r_{j+1}]$ (Fig. 3.18):

$$\hat{X}_h(r) = -\frac{9(r-r_{j+1/3})(r-r_{j+2/3})(r-r_{j+1})}{2(r_{j+1}-r_j)^3} X_j$$
$$+\frac{27(r-r_j)(r-r_{j+2/3})(r-r_{j+1})}{2(r_{j+1}-r_j)^3} X_{j+1/3}$$
$$-\frac{27(r-r_j)(r-r_{j+1/3})(r-r_{j+1})}{2(r_{j+1}-r_j)^3} X_{j+2/3}$$
$$+\frac{9(r-r_j)(r-r_{j+1/3})(r-r_{j+2/3})}{2(r_{j+1}-r_j)^3} X_{j+1} \qquad (3.79)$$

$$\begin{pmatrix} X_h(r) \\ V_h(r) \\ Z_h(r) \end{pmatrix} = \begin{pmatrix} X_j^{\mathrm{R}} \\ V_j^{\mathrm{R}} \\ Z_j^{\mathrm{R}} \end{pmatrix} \frac{2(r-r_{j+1/2})(r-r_{j+1})}{(r_{j+1}-r_j)^2}$$
$$- \begin{pmatrix} X_{j+1/2} \\ V_{j+1/2} \\ Z_{j+1/2} \end{pmatrix} \frac{4(r-r_j)(r-r_{j+1})}{(r_{j+1}-r_j)^2} + \begin{pmatrix} X_{j+1}^{\mathrm{L}} \\ V_{j+1}^{\mathrm{L}} \\ Z_{j+1}^{\mathrm{L}} \end{pmatrix} \frac{2(r-r_j)(r-r_{j+1/2})}{(r_{j+1}-r_j)^2},$$

with the same notation as for (3.69). It is again possible to eliminate intervalwise the three unknowns X_j^{R}, $X_{j+1/2}$, and X_{j+1}^{L} by making the first three moments of (3.59) be zero. The three equations

$$\int_{r_j}^{r_{j+1}} [\hat{X}_h(r) - X_h(r)]\, dr = 0$$
$$\int_{r_j}^{r_{j+1}} [\hat{X}_h(r) - X_h(r)]\, r\, dr = 0 \qquad (3.80)$$
$$\int_{r_j}^{r_{j+1}} [\hat{X}_h(r) - X_h(r)]\, r^2\, dr = 0$$

then lead to the following functional dependence of $X_h(r)$ in the interval $[r_j, r_{j+1}]$

$$X_h(r) = \frac{31X_j + 27X_{j+1/3} - 27X_{j+2/3} + 9X_{j+1}}{40} \; \frac{2(r-r_{j+1/2})(r-r_{j+1})}{(r_{j+1}-r_j)^2}$$
$$-\frac{-X_j + 9X_{j+1/3} + 9X_{j+2/3} - X_{j+1}}{16} \; \frac{4(r-r_j)(r-r_{j+1})}{(r_{j+1}-r_j)^2}$$
$$+\frac{9X_j - 27X_{j+1/3} + 27X_{j+2/3} + 31X_{j+1}}{40} \; \frac{2(r-r_j)(r-r_{j+1/2})}{(r_{j+1}-r_j)^2} \qquad (3.81)$$

3.6.4 Hermite Cubic Elements with Collocation

Non-conforming, non-polluting third-order Hermite elements are characterized by functional dependences of $X_h(r)$, $\hat{X}_h(r)$, $V_h(r)$, and $Z_h(r)$ of the following form:

$$\hat{X}_h(r)=\frac{(r-r_{j+1})^2(3r_j-r_{j+1}-2r)}{(r_j-r_{j+1})^3}\hat{X}_j+\frac{(r-r_j)(r-r_{j+1})^2}{(r_j-r_{j+1})^2}X_j^*$$

$$+\frac{(r-r_j)^2(3r_{j+1}-r_j-2r)}{(r_{j+1}-r_j)^3}\hat{X}_{j+1}+\frac{(r-r_{j+1})(r-r_j)^2}{(r_{j+1}-r_j)^2}X_{j+1}^* \ , \tag{3.82}$$

$$\begin{pmatrix}X_h(r)\\V_h(r)\\Z_h(r)\end{pmatrix}=\begin{pmatrix}X_j\\V_j\\Z_j\end{pmatrix}\frac{2(r-r_{j+1/2})(r-r_{j+1})}{(r_{j+1}-r_j)^2}$$

$$-\begin{pmatrix}X_{j+1/2}\\V_{j+1/2}\\Z_{j+1/2}\end{pmatrix}\frac{4(r-r_j)(r-r_{j+1})}{(r_{j+1}-r_j)^2}+\begin{pmatrix}X_{j+1}\\V_{j+1}\\Z_{j+1}\end{pmatrix}\frac{2(r-r_j)(r-r_{j+1/2})}{(r_{j+1}-r_j)^2} \ .$$

If we impose the three moments (3.80) to express the nodal points of X_h by those of $\hat{X}_h$, we obtain a system which does not permit explicit elimination of the two nodal values at the edge. As a consequence, we change the method by replacing (3.59) by two collocation conditions

$$\begin{aligned}X_j&=\hat{X}_j\\X_{j+1}&=\hat{X}_{j+1} \ ,\end{aligned}\tag{3.83}$$

and by satisfying

$$\int_{r_j}^{r_{j+1}}[\hat{X}_h(r)-X_h(r)]\,dr=0 \ . \tag{3.84}$$

This latter condition (3.84) enables us to express the mid-point nodal value $X_{j+1/2}$ in terms of the unknowns of $\hat{X}_h$:

$$X_{j+1/2}=\frac{X_j+X_{j+1}}{2}+\frac{r_{j+1}-r_j}{8}(X_j^*-X_{j+1}^*) \ . \tag{3.85}$$

3.7 Applications and Comparison Studies (with M.-A. Secrétan)

We have modified our 1D MHD stability code THALIA (*Appert* et al. 1975b) such that we can apply all the methods presented in Sects. 3.5, 6 to the test cases.

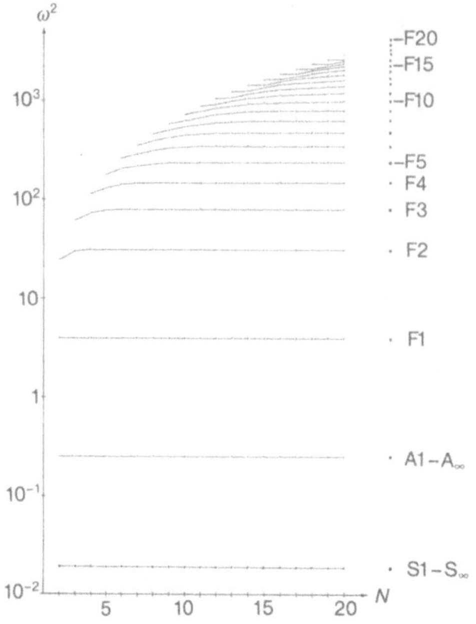

Fig. 3.20. The spectrum of test case A calculated with the unpolluted, conforming, lowest order finite element approach

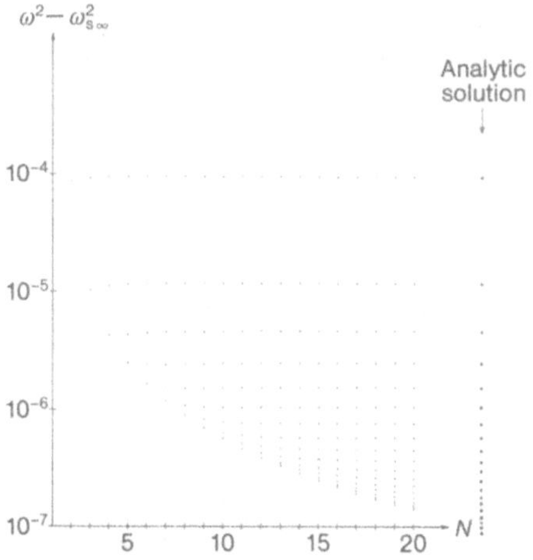

Fig. 3.21. The Slow mode spectrum of test case A calculated with the unpolluted, conforming, lowest order finite element approach

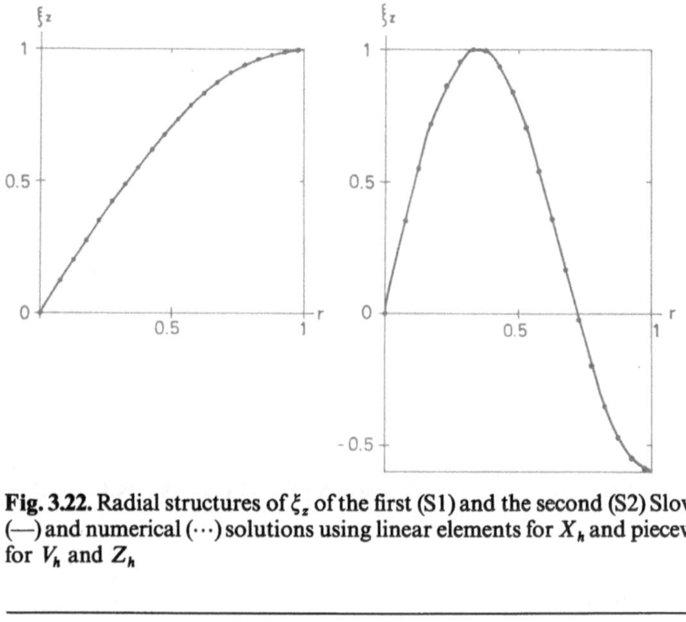

Fig. 3.22. Radial structures of ξ_z of the first (S1) and the second (S2) Slow mode of the analytic (—) and numerical (···) solutions using linear elements for X_h and piecewise constant elements for V_h and Z_h

3.7.1 Application to Test Case A

We apply our modified code to the test case A (3.35) using elements (3.66) for the three vector components. The numerical spectrum as a function of N is shown in Fig. 3.20. We find $N-1$ degenerate eigenvalues $\omega^2 = k^2 = 0.25$. This means that all the Alfvén modes are degenerate. The eigenmodes of this class have a δ-function-like character as expected. The pollution, which occurs when standard linear finite elements for all the three components are used, has now disappeared. As before, the Fast modes are well represented. The Slow wave spectrum is studied by again magnifying the part of the spectrum around $\omega^2 = \gamma p k^2/(1+\gamma p)$ (Fig. 3.21). With our new "ecological" finite element approach we have no problems detecting and numbering each of the modes (Table 3.1, p. 49). The newly created modes which appear with increasing N, are in correct order and never cross each other as they do in the polluted case (Fig. 3.14). The eigenmodes are very close to what one expects from the analytic calculation. This can be seen in Fig. 3.22, where the numerically calculated first and second Slow modes (dots) are compared with the analytic solutions. This figure has to be compared with the result from the polluted approach in Fig. 3.15.

What we can learn from this is that imposing regularities on a solution can have catastrophic consequences. In our test case, when using linear elements for all three components, the approximated operator did not converge towards the physical one. A well-represented spectrum is only obtained by reducing regularity for V_h. Linear finite elements for X_h and piecewise constant elements for V_h and Z_h lead to an approximated operator which converges towards the physical one (see also Chap. 1). The same result is obtained when applying the methods described in Sects. 3.5.2–4 and Sects. 3.6.1–4 to the test case A.

3.7.2 Application to Test Case B

In this section we try to give some hints as to how a continuous spectrum can be detected numerically. For this purpose we choose test case B with a density profile defined in (3.36) and $\varepsilon = 0.1$. We again apply the "ecological" lowest-order conforming elements described in Sect. 3.5.1. We first study in Fig. 3.23 how the $1/(r-r_A)$ "singularity" of the dominant poloidal component V of the Alfvén mode develops when the number of radial intervals N is increased. The position r_A corresponds to $\omega_A^2 = k^2/(1-\varepsilon r_A^2)$. For all N, the modes are normalized numerically by

$$\int_0^{r_p} \xi_h^2 r\, dr = 1 \quad . \tag{3.86}$$

The $1/(r-r_A)$ behavior of V can clearly be recognized. Its maximum amplitude increases with $N^{1/2}$ and its half-width decreases with $1/N$. The fact that the amplitude only grows with $N^{1/2}$ and not with N is due to the normalization (3.86) which does not represent the truth. In reality, such a singular solution is

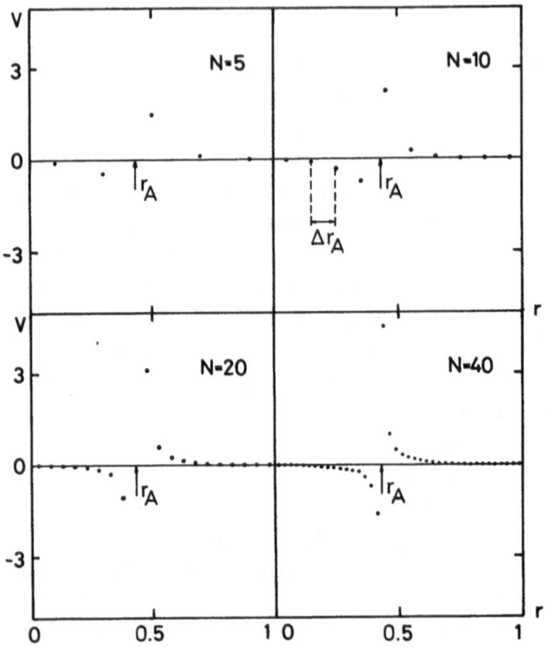

Fig. 3.23. Numerically calculated singular solutions of the Alfvén continuum: $\varepsilon=0.1$. The number of intervals used are 5, 10, 20, and 40. Only the dominant component V is given. r_A denotes the analytically determined position of the singularity and Δr_A is the uncertainty with which this position can be predicted numerically. (From *Appert* et al. 1975a)

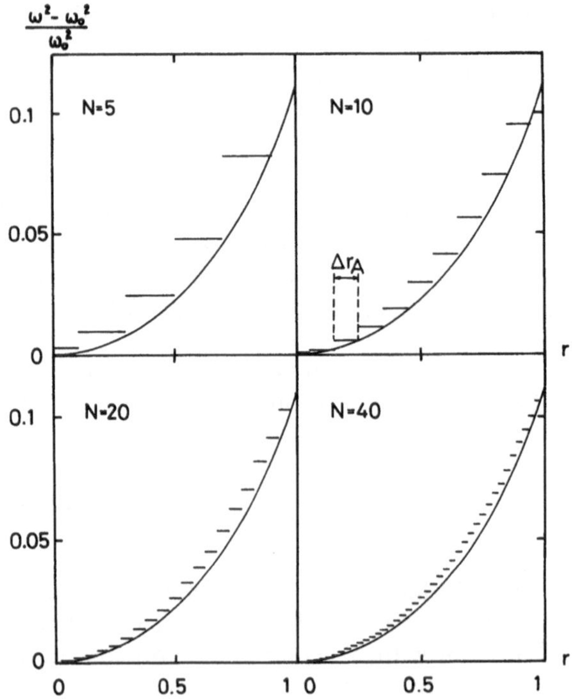

Fig. 3.24. Numerical detection of the continuum of the Alfvén modes. The solid lines denote the analytically calculated positions of the singularity of the modes corresponding to eigenmodes ω^2 lying in continuum. Δr_A is the uncertainty with which these positions can be found numerically. (From *Appert* et al. 1975a)

not normalizable, since

$$\int_0^{r_P} \xi^2 r\,dr \propto \int_0^{r_P} \frac{r\,dr}{(r-r_A)^2} = \infty \ . \qquad (3.87)$$

How well can we predict the position of the singularity $r=r_A$ numerically? When studying the radial behavior of V in Fig. 3.23, we can predict that $r=r_A$ should lie between the maxima and the minima of the modes. This radial uncertainty Δr_A in the prediction of r_A is used in Fig. 3.24, where we show the Alfvén frequency $\omega^2(r)=k^2/(1-\varepsilon r^2)$ as a function of r for $N=5$, 10, 20, and 40. The solid line represents the analytic solution. The numerical results include the uncertainties Δr_A. We find a good correspondence between the analytic and numerical results. From Figs. 3.23, 24 we conclude that a numerical spectrum may be regarded as continuous in a certain frequency band, if

- the associated normalized modes develop singularities with increasing resolution;
- the number of modes in a fixed frequency interval increases with increasing number of intervals;
- the uncertainty region of the singularities converges towards the frequency polygon associated with the continuous spectrum when the resolution is increased.

3.7.3 Application to Test Case C

We now apply each of our unpolluting conforming and non-conforming methods to the test case C (Sect. 3.2.3) which deals with a global Bessel-function-like eigenmode (Fig. 3.6). To be able to quantify the power of the different methods, convergence studies are performed. As stated in Chap. 1, the convergence behavior of the eigenvalue ω^2 of a finite element approach, using basis functions of order l, is expected to be $O(h^{2l})$ or $O(1/N^{2l})$, where h is the mesh size and N the number of intervals. The radial component X_h of the eigenfunction converges in $O(h^{l+1})$, its derivative dX_h/dr and the two other components V_h and Z_h in $O(h^l)$. In all of the eight convergence studies an equidistant mesh has been used.

In Fig. 3.25 the eigenvalues ω^2 are plotted as a function of the number N of mesh cells (in a $1/N^2$ scale) for the conforming (L1) and the non-conforming (L−1) linear ($l=1$) finite elements. Both methods converge quadratically towards $\omega^2=-9.017\times10^{-4}$. The gradient of the convergence line is slightly smaller for non-conforming than for conforming elements. Both methods converge from "above", i.e., from higher ω^2 towards smaller ω^2 with increasing N.

For quadratic conforming (L2) and non-conforming (L−2) elements, the convergence curves $\omega^2(N)$ are shown in Fig. 3.26. For the conforming elements, the straight line towards the value $\omega^2=-9.018\times10^{-4}$ indicates a convergence

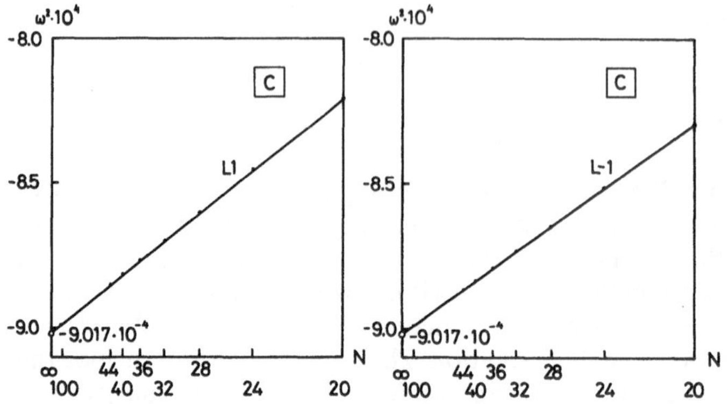

Fig. 3.25. Convergence curves $\omega^2(N)$ for test case C using conforming (L1) and non-conforming hybrid (L -1) linear elements. Both methods converge quadratically towards $\omega^2 = -9.017\,10^{-4}$

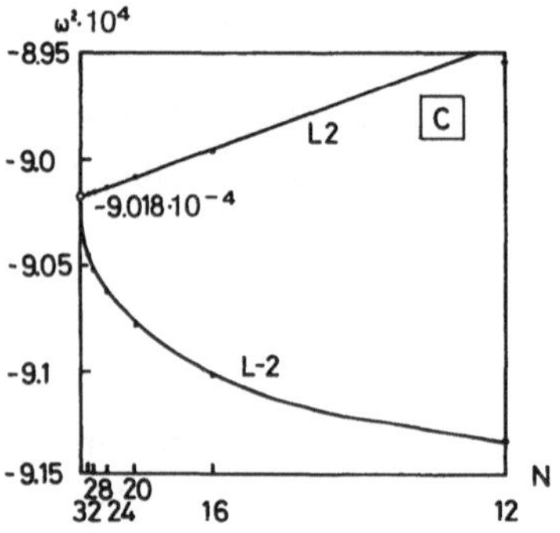

Fig. 3.26. Convergence curves $\omega^2(N)$ for test case C using conforming (L2) and non-conforming hybrid (L-2) quadratic elements. The conforming approach shows a convergence in $O(h^4)$

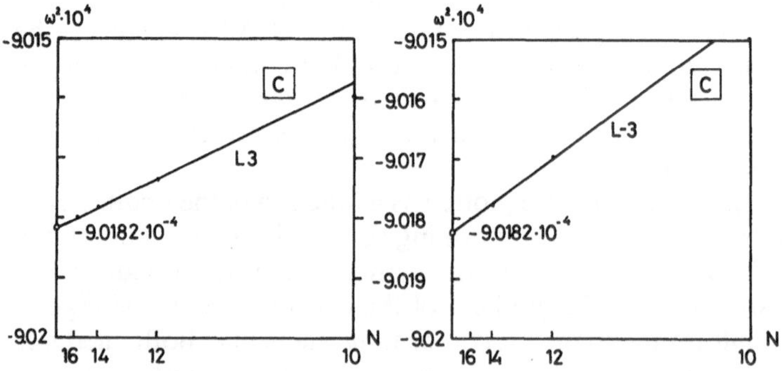

Fig. 3.27. $\omega^2(N)$ for test case C using conforming (L3) and non-conforming hybrid (L -3) cubic Lagrange elements. Both methods converge in $O(h^6)$

in $O(h^4)$. The non-conforming elements do not seem to follow the expected convergence behavior, at least in the range of $N<30$. We also have to note that the eigenvalues of the non-conforming approach lie below the physical eigenvalue.

The convergence curves for the conforming (L3) and non-conforming Lagrange cubic finite element approaches are shown in Fig. 3.27. Both methods converge from "above" towards $\omega^2=-9.0182\times10^{-4}$, the non-conforming approach having a slightly bigger gradient.

Finally, the cubic Hermite elements lead to convergence curves as presented in Fig. 3.28. The conforming approach (H3) tends towards $\omega^2=-9.0188\times10^{-4}$ whereas the non-conforming approach (H−3) with collocation tends towards $\omega^2=-9.015\times10^{-4}$. The collocation method therefore strongly stabilizes the eigenvalue.

If we wish to compare the power of these methods, the number of intervals is not a good parameter. What we have to do is to compare results which cost the same price on a given computer. This can be achieved by a price-performance diagram such as that given in Fig. 3.29. Here we plot the absolute error of the

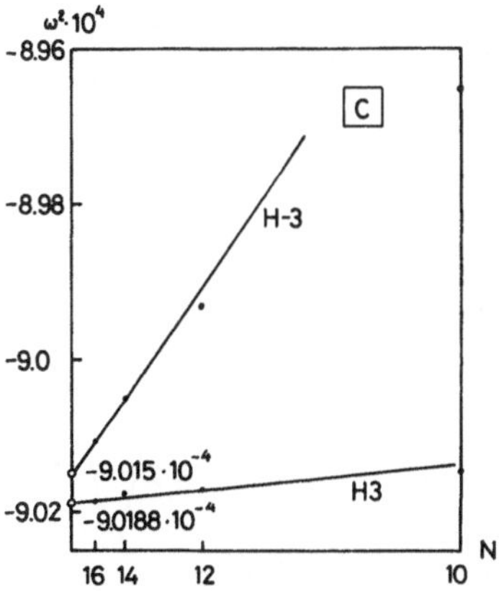

Fig. 3.28. $\omega^2(N)$ for test case C using conforming (H3) and non-conforming hybrid (with collocation) (H−3) cubic Hermite elements. Both methods converge in $O(h^6)$

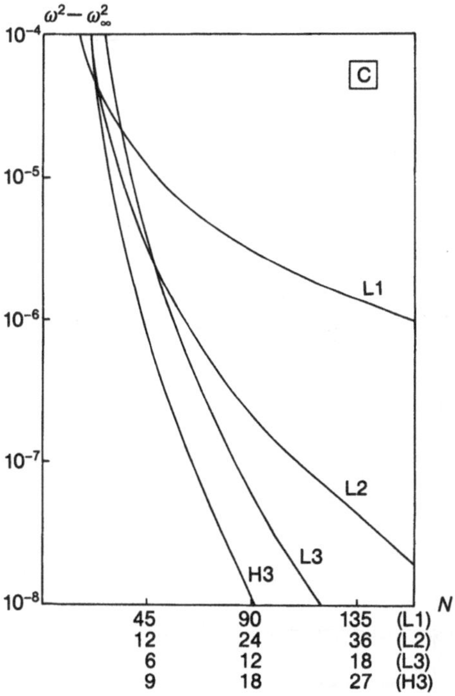

Fig. 3.29. Price-performance diagram $\omega^2(N)-\omega^2(\infty)$ for test case C comparing linear (L1), quadratic (L2), cubic Lagrange (L3), and cubic Hermite (H3) elements. The same values of the abcissa correspond to the same computing costs on a vector machine. Hermite elements give the best results

eigenvalue as a function of the number of intervals N, choosing different N values for different methods in such a way that the computing times are comparable. For instance, on a vector machine when representing the eigenmode by linear elements the CPU time taken for 45 intervals is about the same as taking 12 intervals for quadratic elements, 6 intervals for Lagrange-cubic or 9 intervals for Hermite-cubic elements. We see in Fig. 3.29 that Hermite elements are the most powerful ones. Note that for a scalar machine, such a

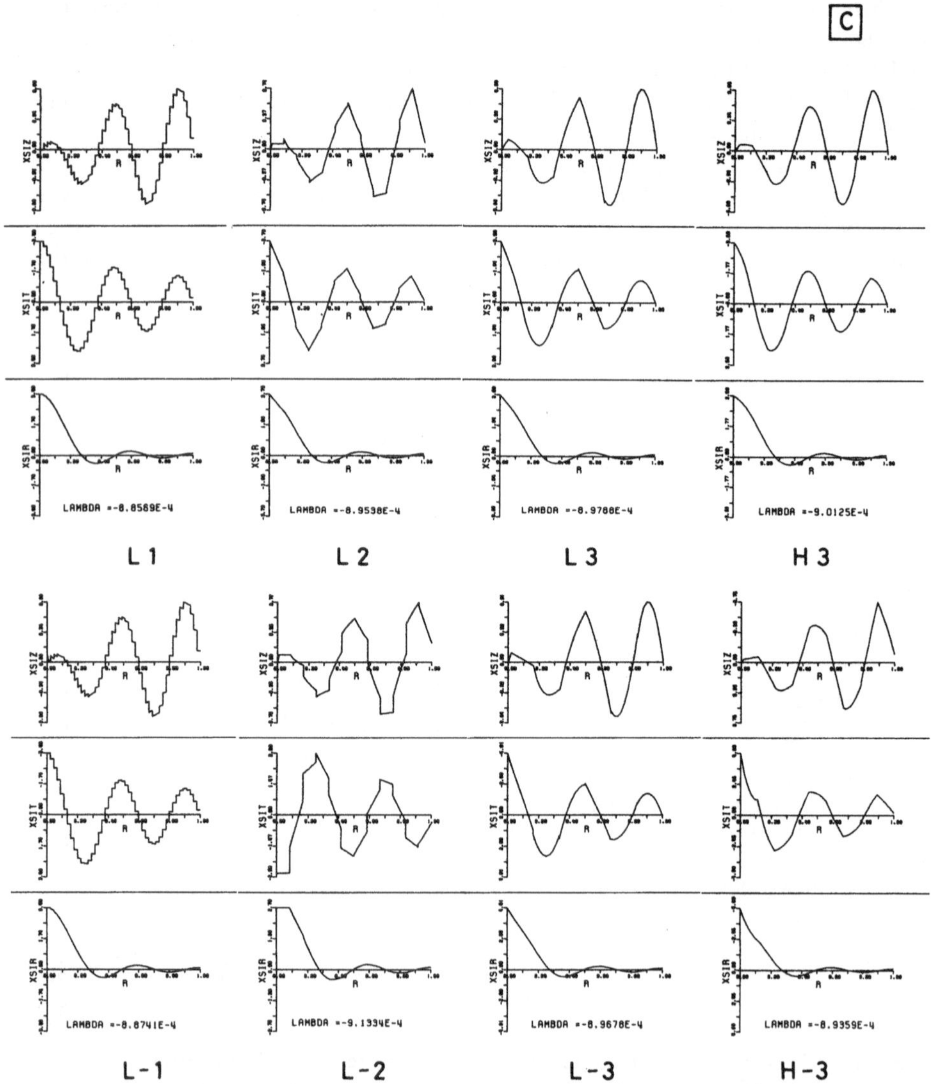

Fig. 3.30. The three vector components of the eigenmode of test case C for 8 finite element methods

comparison of methods gives a more advantageous result for the lowest-order elements. To finish the price-performance comparison, we represent in Fig. 3.30 the three components $\xi_z(r)$, $\xi_\theta(r)$, $\xi_r(\theta)$ of the eigenmodes using $N_1=45$ (for L1 and L−1), $N_2=12$ (for L2 and L−2), $N_3=6$ (for L3 and L−3), and $N_4=9$ (for H3 and H−3) intervals. The calculation of each of these 8 modes costs about the same price on a vector machine.

3.7.4 Application to Test Case F

Finally, we apply our conforming and non-conforming methods to test case F (Sect. 3.2.6) which deals with a localized internal kink mode (Fig. 3.9). Again, we first study the convergence curves for conforming (L1) and non-conforming hybrid (L−1) elements. We see in Fig. 3.31 that both methods converge towards $\omega^2=-5.87\times10^{-5}$ but from opposite sides. Using conforming elements, an unstable mode can be found for $N>26$. If, for instance, we had stopped the calculations with $N=24$ we would have missed this instability. With hybrid elements one finds very unstable modes with only a few intervals. A convergence study must be done to find out which of the obtained modes are physical ones. In Fig. 3.31 we have also plotted the eigenvalue (denoted by ×) of the second most unstable mode which tends towards zero as the resolution is

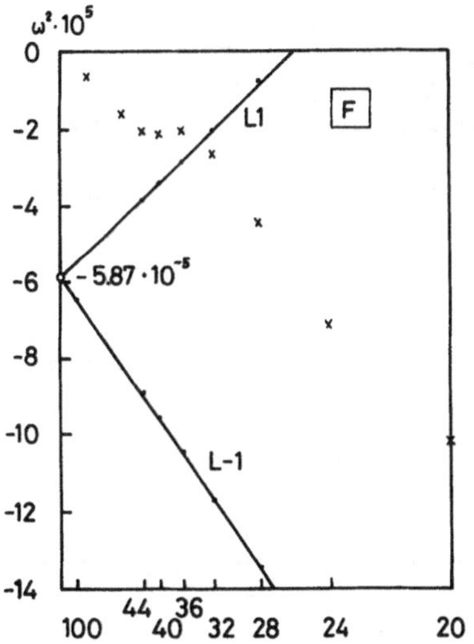

◄ **Fig. 3.31**

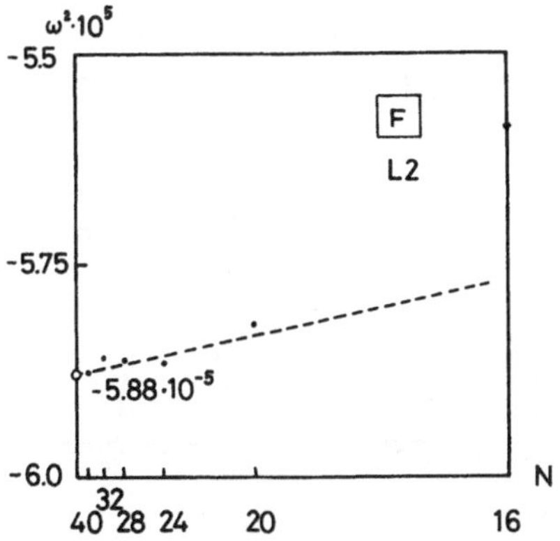

Fig. 3.31. Convergence studies for test case F using conforming (L1) and non-conforming hybrid (L−1) linear elements. Both methods converge quadratically towards $\omega^2=-5.87\,10^{-5}$, the conforming approach from "above", the non-conforming approach from "below"

Fig. 3.32. $\omega^2(N)$ for test case F using conforming quadratic elements (L2). Quartic convergence towards $\omega^2=-5.88\,10^{-5}$ is observed. The non-conforming approach does not show any convergence law

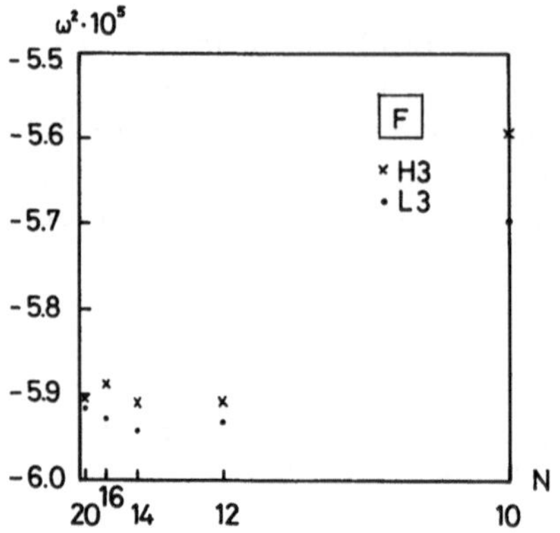

Fig. 3.33. $\omega^2(N)$ for test case F using conforming cubic Lagrange (L3) and Hermite (H3) elements. No evident convergence law can be seen in the range $N \leq 20$

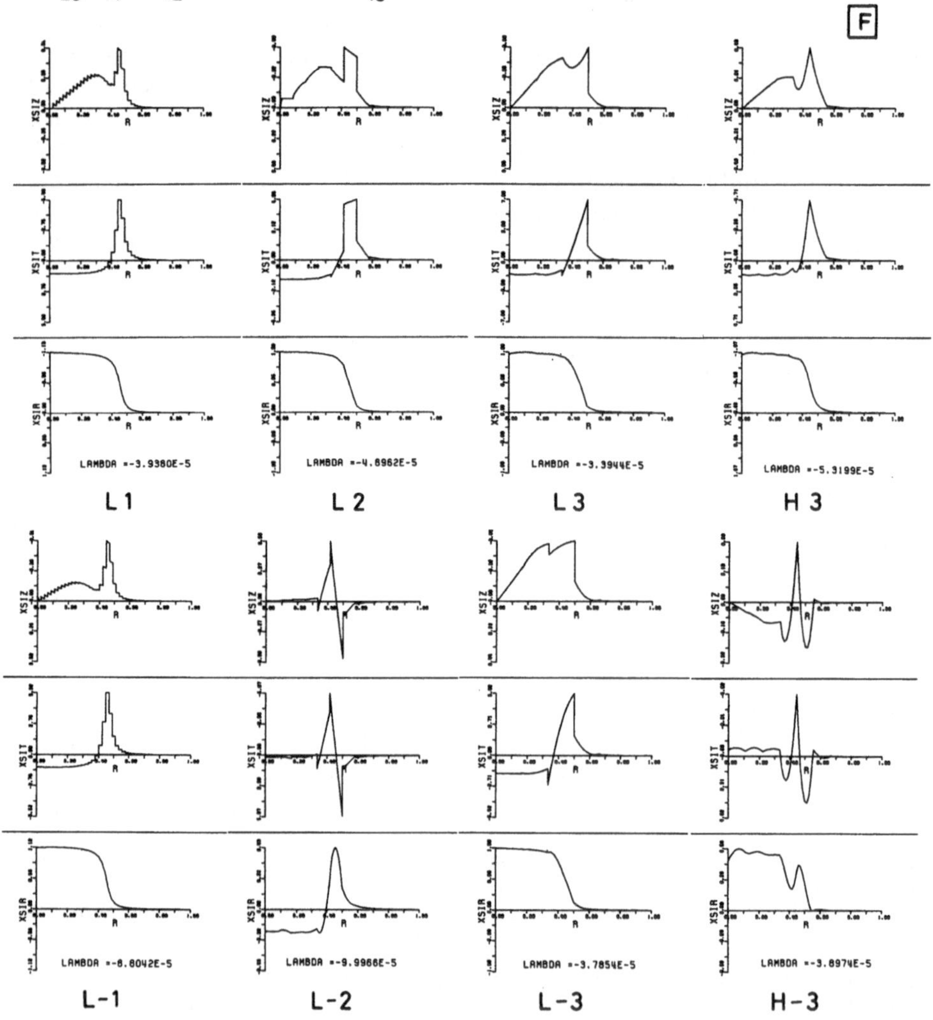

Fig. 3.34. The three components of the eigenmode of test case F for the 8 finite element methods

increased. The structure of this second unstable eigenmode is very close to a δ-function. As the resolution is increased, this mode tends towards one belonging to the continuous spectrum. This means that with non-conforming elements the continuous spectrum can be destabilized.

The convergence curve for conforming (L2) quadratic elements is shown in Fig. 3.32. Here we must note that the scale of the ω^2-axis is 20 times smaller than that in Fig. 3.31. The convergence seems to be $O(h^4)$ with small oscillations around the convergence line. We remember that when representing the global mode, the convergence behavior for non-conforming quadratic elements (Fig. 3.26) was unclear. Now, when one applies this approach to the internal kink, it becomes so catastrophic that we do not show the results.

Conforming cubic Lagrange (L3) and Hermite (H3) elements show convergence behaviors as shown in Fig. 3.33. No clear convergence in $O(h^6)$ can be recognized for $N \leqq 20$. Non-conforming elements have even worse convergence properties.

With all these results, it is not necessary to make a price-performance diagram as given for a global mode in Fig. 3.29. It is evident that the conforming and non-conforming linear elements give the best results. This conclusion is underlined when one looks at the eigenmodes shown in Fig. 3.34, again using the number of intervals corresponding to the same CPU costs on a vector machine. One rapidly realizes the superiority of linear elements.

3.8 Discussion and Conclusion

Before choosing a numerical method to solve a physical problem, several considerations have to be made:

- The chosen method has to converge to the physical problem.
- The method has to be precise so that the computing costs remain small.
- The method should lead to easy coding.

All eight methods discussed in Sects. 3.5 and 6 are unpolluted and thus satisfy the first point. For the test case C discussed in Sect. 3.7.3 the best precision for given computing costs is obtained with cubic-Hermite polynomials as basis functions. However, when the eigenmode is not further global but has a local character as in the case of an internal kink mode discussed in Sect. 3.7.4, the least regular elements give the best results. Increasing regularity does not help in the representation of localized or jumping modes.

The third point opts for a linear finite element approach. The functional dependence in one interval is simple, and Simpson's rule can be used to fulfil the consistency requirement. Cubic finite elements lead to a fairly complicated functional dependence in each interval. The integration scheme has to be precise, at least up to seventh-order polynomials, in order to remain consistent with the choice of basis functions. Here, consistency means that at least one constant can be exactly reproduced. Coding for cubic elements is evidently more complicated than for linear elements. However, a modular organization of the code does not increase the complexity of coding too much.

As a final conclusion we can say that the third-order Hermite polynomials give the best results for global modes, but if a spectrum includes localized solutions, the least regular basis functions are more advantageous.

4. Two-Dimensional Finite Elements Applied to Cylindrical Geometry

Before treating realistic cases in complicated toroidal and helical geometries, we first apply the two-dimensional finite element approaches to cylindrical geometry, the results of which can be compared to the one-dimensional calculations reported in Chap. 3.

4.1 Conforming Finite Elements

In the variational form (3.44–49) a normal mode analysis has been performed in θ and z (3.2). This was possible because all equilibrium quantities do not vary in these directions. We now consider the same geometry, but we only perform a normal mode analysis in z direction and resolve the $\exp(im\theta)$ variation of the mode numerically. As in Chap. 3 we introduce the new variables

$$X = r\xi_r$$
$$V = (B_z\xi_\theta - B_\theta\xi_z)/B_z \tag{4.1}$$
$$Z = r\xi_z/B_z \ .$$

In all calculations in this section we only consider the fixed boundary case (3.11), i.e., we always impose for all θ:

$$X(r_p, \theta) = 0 \ . \tag{4.2}$$

The variational form is (2.23–26)

$$\delta\mathscr{L} = \delta(W_p - \omega^2 K) = 0 \ , \quad \text{with} \tag{4.3}$$

$$W_p = \frac{1}{2}\int_0^{r_p}\int_0^{2\pi}\frac{dr}{r}d\theta\left[\frac{B_\theta^2}{r^2}\left|\frac{\partial X}{\partial\theta}+ikqX\right|^2 + B_z^2\left|\frac{\partial X}{\partial r}+\frac{\partial V}{\partial\theta}\right|^2\right.$$
$$+B_\theta^2\left|\frac{\partial X}{\partial r}-ikqV-\frac{2}{r}X\right|^2+\gamma p\left|\frac{\partial X}{\partial r}+\frac{\partial V}{\partial\theta}+\frac{B_\theta}{r}\left(\frac{\partial Z}{\partial\theta}+ikqZ\right)\right|^2$$
$$\left.-2\frac{B_\theta}{r^2}\frac{d}{dr}(rB_\theta)\,|X|^2\right] \tag{4.4}$$

$$K = \frac{1}{2}\int_0^{r_p}\int_0^{2\pi}\frac{\varrho\, dr}{r}d\theta\{|X|^2 + |rV + B_\theta Z|^2 + B_z^2|Z|^2\} \ .$$

Note that expressions (4.4) are the same as (3.44 and 48) if m is replaced by $\partial/\partial\theta$. The variables X, V, and Z are now complex:

$$\begin{aligned} X(r,\theta) &= X^{\mathrm{R}}(r,\theta) + \mathrm{i}X^{\mathrm{I}}(r,\theta) \\ V(r,\theta) &= V^{\mathrm{R}}(r,\theta) + \mathrm{i}V^{\mathrm{I}}(r,\theta) \\ Z(r,\theta) &= Z^{\mathrm{R}}(r,\theta) + \mathrm{i}Z^{\mathrm{I}}(r,\theta) \ . \end{aligned} \tag{4.5}$$

The upper indices R and I denote the real and imaginary parts of the components.

Let U be the set of all θ periodic functions $\boldsymbol{u}=(X, V, Z)$ in the domain $\Omega=\{0<r<r_{\mathrm{p}}, 0<\theta<2\pi\}$ which fulfil some regularities such that the integrals (4.4) have a sense [mathematical details can be found in *Jaccard* (1980) and *Evequoz* (1980)]. The variational formulation (4.3 and 4) is then:

"Find real numbers ω^2 and nontrivial $\boldsymbol{u}=(X, V, Z)\in U$ with $X(r_p,\theta)=0$ for all θ satisfying $\delta\mathscr{L}=\delta(W_p-\omega^2 K)=0$." (4.6)

If U_h is a finite-dimensional subspace of U, the corresponding approximate problem is

"Find real numbers ω^2 and non-trivial functions $\boldsymbol{u}_h=(X_h, V_h, Z_h)\in U_h$ with $X_h(r_p,\theta)=0$ for all θ such that $\delta\mathscr{L}_h=\delta(W_{\mathrm{p}}-\omega^2 K)_h=0$." (4.7)

4.1.1 Conforming Triangular Finite Elements

As a first approach to solving Problem (4.7) we apply classical triangular finite elements. The domain $\Omega=\{0<r<r_{\mathrm{p}}, 0<\theta<2\pi\}$ is subdivided into $N=N_r * N_\theta$ intervals. Each rectangular cell is cut into two triangles, as shown in Fig. 4.1. All vector components X_h, V_h, and Z_h are chosen to be continuous in Ω, (in the case of triangular elements, X_h, V_h, and Z_h are polynomials of degree 1 in r and θ) and have to satisfy the periodicity and boundary conditions. If $\tilde{U}_h$ is the subspace of all functions $\boldsymbol{u}_h=(X_h, V_h, Z_h)$ such that

$$\begin{aligned} &\frac{\partial X_h}{\partial r}+\frac{\partial V_h}{\partial \theta}=0 \\ &Z_h=0 \ , \end{aligned} \tag{4.8}$$

then $\tilde{U}_h$ includes the Alfvén modes of the analytic solution (Sect. 3.1.3). We can prove that this approach leads to a space $\tilde{U}_h$ of dimension 1. In fact, on each triangle, X_h and V_h have the form

$$\begin{aligned} X_h(r,\theta) &= a_1+b_1 r+c_1\theta \\ V_h(r,\theta) &= a_2+b_2 r+c_2\theta \ , \end{aligned} \tag{4.9}$$

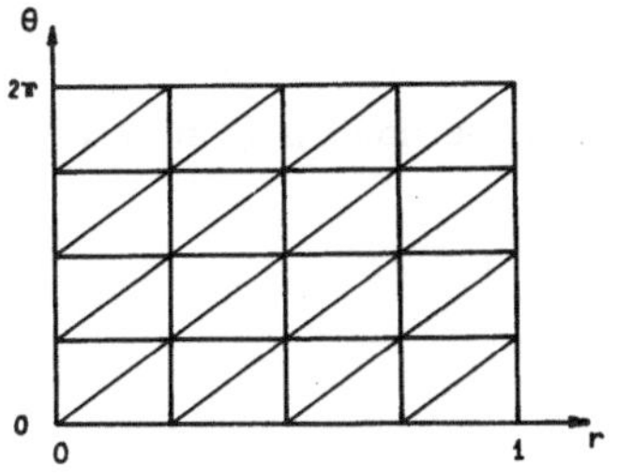

Fig. 4.1. Triangularization of the (r, θ) domain

and (4.8) is satisfied only if

$$b_1+c_2=0 \ . \tag{4.10}$$

Since there are $2*N$ triangles, we have to satisfy $2*N$ such conditions for the first equation of (4.8). It is easy to see that we have $2*N$ degrees of freedom for X_h and V_h [$2*(N_r+1)*(N_\theta+1)$ unknowns, minus $2*(N_r+1)$ periodicity conditions, minus $2*N_\theta$ boundary conditions]. As a consequence, there is only one mode in $\tilde{U}_h$. Calculations have shown that this approach is a polluted one and is therefore inappropriate.

4.1.2 Conforming Lowest-Order Quadrangular Finite Elements

For the two-dimensional problem (4.7), the conditions which prevent spectrum pollution have been formulated mathematically by *Jaccard* and *Evequoz* (1981). They have shown that to avoid pollution, a sufficient condition is that the basis functions have a radial dependence in S_{p+1}^{l+1} (Sect. 1.4) for X_h and S_p^l for V_h with $p \geq 2l \geq 0$. There is no restriction in the poloidal direction. However, to satisfy incompressibility condition

$$\frac{\partial X}{\partial r}+\frac{\partial V}{\partial \theta}=0 \tag{4.11}$$

"as precisely as possible", the polynomial variation of the basis functions of V_h should be one order higher than that for X_h.

The domain Ω is subdivided into $N=N_r*N_\theta$ mesh cells as shown in Fig. 4.2. For X_h we take basis functions in $S_1^1(r)\otimes S_1^1(\theta)$; for V_h, basis functions in $S_0^0(r)\otimes S_2^1(\theta)$; and those in $S_0^0(r)\otimes S_1^1(\theta)$ for Z_h (*Berger* et al. 1976). The approximation is then written as;

$$\begin{aligned}
X_h(r,\theta) &= \sum_{i=1}^{N_r-1}\sum_{j=0}^{N_\theta} X_{ij}\, e_i(r)\, e_j(\theta) \\
V_h(r,\theta) &= \sum_{i=1}^{N_r} c_{i-1/2}(r)\left[\sum_{j=0}^{N_\theta} V_{i-1/2,j} f_j(\theta)+\sum_{j=1}^{N_\theta} V_{i-1/2,j-1/2} f_{j-1/2}(\theta)\right] \\
Z_h(r,\theta) &= \sum_{i=1}^{N_r} c_{i-1/2}(r)\sum_{j=0}^{N_\theta} Z_{i-1/2,j}\, e_j(\theta) \ ,
\end{aligned} \tag{4.12}$$

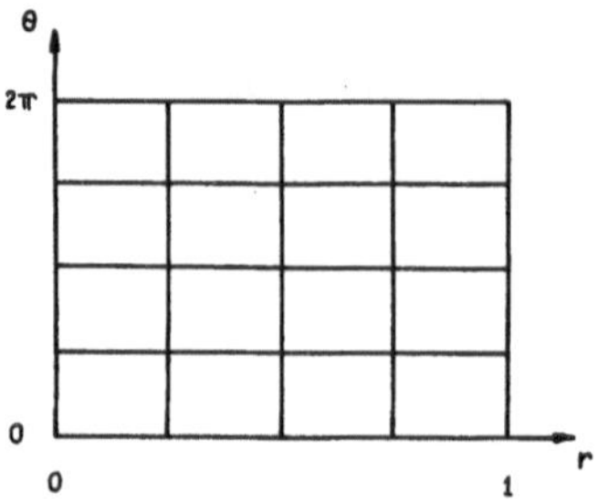

Fig. 4.2. Discretization of the domain Ω in $N_r * N_\theta$ rectangles

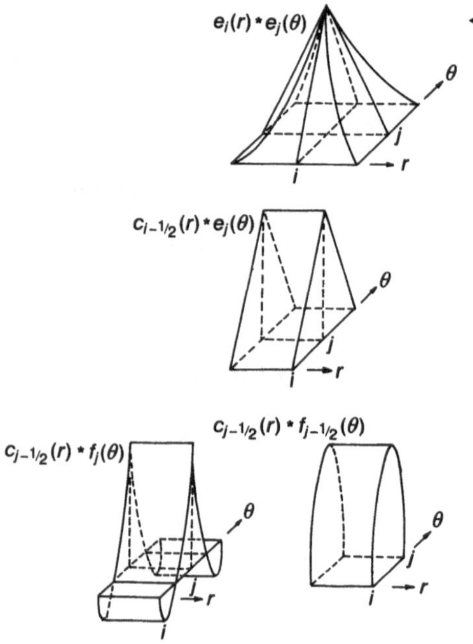

◀ **Fig. 4.3.** Shape of the basis functions $e_i(r) * e_j(\theta)$, $c_{i-1/2}(r) * e_j(\theta)$, $c_{i-1/2}(r) * f_j(\theta)$, and $c_{i-1/2}(r) * f_{j-1/2}(\theta)$

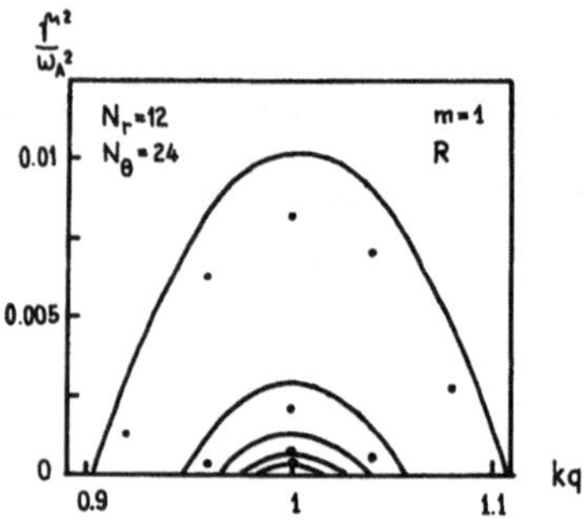

Fig. 4.4. The first five unstable modes of test case D (—). (···) are the eigenvalues obtained with a *conforming finite element* approach using $N_r = 12$ and $N_\theta = 24$ intervals. (From *Gruber* 1978)

where X_{ij}, $V_{i-1/2,j}$, and $Z_{i-1/2,j}$ are complex coefficients restricted by the periodicity conditions

$$\begin{aligned} X_{i,N_\theta+1} &= X_{i1} \\ V_{i-1/2,N_\theta+1} &= V_{i-1/2,1} \\ Z_{i-1/2,N_\theta+1} &= Z_{i-1/2,1} \,. \end{aligned} \tag{4.13}$$

The basis functions $c_{k-1/2}(x)$, $e_k(x)$, $f_k(x)$, and $f_{k-1/2}(x)$ are the piecewise constant, linear and quadratic functions defined in Chap. 1. The two-dimensional functions $e_i(r)\,e_j(\theta)$, $c_{i-1/2}(r) f_j(\theta)$, $c_{i-1/2}(r) f_{j-1/2}(\theta)$, and $c_{i-1/2}(r)\,e_j(\theta)$ are shown in Fig. 4.3.

We first apply this approach to test case A; a homogeneous currentless plasma cylinder for which the Alfvén spectrum is infinitely degenerate for $\omega^2 = k^2$ and for which the eigenmode has to fulfil (4.8). We find that the Alfvén modes have the right degeneracy which means that this approach is well suited for test case A.

The same method is applied to test case D (Sect. 3.2.4). The results are presented in Fig. 4.4. The solid curves representing the unstable region for $m=1$, $k=-0.2$ are the same as those given in Fig. 3.7. The small circles are the results obtained with the two-dimensional finite element approach using $N_r = 12$ and $N_\theta = 24$ intervals. We find that this finite element approach strongly stabilizes the most unstable modes and that this stabilization effect is mainly due to insufficient resolution in the poloidal direction.

This stabilizing effect is still more pronounced when we try to calculate the unstable region for test case E (Fig. 3.8). No unstable mode is found with $N_r = 12$ and $N_\theta = 40$ for the finite elements given in (4.12). Here again, many more θ intervals have to be taken in order to find any unstable modes.

Even though the finite element approach (4.12) is not polluted, the results obtained are unsatisfactory. Too many poloidal intervals have to be taken in order to predict instability for test case E. Where does this stabilization come from?

While describing the test cases D and E in Sect. 3.2, we have seen that the unstable region is found for cases for which the quantity $F = (kB_z + m/rB_\theta) = B_\theta/r(kq+m)$ is near zero. Especially for $F=0$ there is an infinite number of unstable modes. In the 1D calculations it is evident that $F=0$ can be fulfilled identically. In the 2D calculation, however, F is an operator, and the condition that $F=0$ be satisfied for integer values of $kq=-m$ now becomes

$$\begin{aligned} &\frac{\partial X}{\partial \theta} - imX = 0 \\ &\frac{\partial V}{\partial \theta} - imV = 0 \ . \end{aligned} \tag{4.14}$$

The impossibility of fulfilling these conditions gives the strong stabilization effect that we observe.

This problem of precision can be tackled by Fourier-analysing in the poloidal angle, instead of by using finite elements. This is the approach used in PEST (*Grimm* et al. 1976) with which the 1D results can be reproduced. Our approach is to choose so-called finite hybrid elements (*Gruber* 1978), which are equivalent to non-conforming finite elements. These are presented in the following section.

4.2 Non-Conforming, Finite Hybrid Elements

4.2.1 Finite Hybrid Elements Formulation

Consider the space U defined by the set of functions $\boldsymbol{u}=(X, X^{(1)}, X^{(2)}, X^{(3)}, V, V^{(2)}, Z, Z^{(2)})$ such that:

$$X, \frac{\partial X}{\partial r}, \frac{\partial X}{\partial \theta} \quad \text{are "sufficiently regular"}^{1} \text{ and}$$
$$X(0,\theta)=X(r_p,\theta)=0 \quad \text{for all} \quad \theta \ , \tag{4.15}$$

$$X^{(1)}, \frac{\partial X^{(1)}}{\partial \theta} \quad \text{are "sufficiently regular" and}$$
$$\int_{\Omega} (X^{(1)}-X)\eta \, dr \, d\theta=0 \quad \text{for all} \quad \eta \in L^2(\Omega) \tag{4.16}$$

$$X^{(2)} \text{ is "sufficiently regular" and}$$
$$\int_{\Omega} (X^{(2)}-X)\eta \, dr \, d\theta=0 \quad \text{for all} \quad \eta \in L^2(\Omega) \tag{4.17}$$

$$X^{(3)}, \frac{\partial X^{(3)}}{\partial r} \quad \text{are "sufficiently regular" and}$$
$$\int_{\Omega} (X^{(3)}-X)\eta \, dr \, d\theta=0 \quad \text{for all} \quad \eta \in L^2(\Omega) \tag{4.18}$$

$$V, \frac{\partial V}{\partial \theta} \quad \text{are "sufficiently regular"} \tag{4.19}$$

$$V^{(2)} \text{ is "sufficiently regular" and}$$
$$\int_{\Omega} (V^{(2)}-V)\eta \, dr \, d\theta=0 \quad \text{for all} \quad \eta \in L^2(\Omega) \tag{4.20}$$

$$Z, \frac{\partial Z}{\partial \theta} \quad \text{are "sufficiently regular"} \tag{4.21}$$

$$Z^{(2)} \text{ is "sufficiently regular" and}$$
$$\int_{\Omega} (Z^{(2)}-Z)\eta \, dr \, d\theta=0 \quad \text{for all} \quad \eta \in L^2(\Omega) \ , \tag{4.22}$$

and $\boldsymbol{u}(r,0)=\boldsymbol{u}(r,2\pi)$. The integral conditions in (4.16–18, 20 and 22) imply that

$$X^{(1)}=X^{(2)}=X^{(3)}=X \ , \quad V^{(2)}=V, \quad Z^{(2)}=Z \ . \tag{4.23}$$

1 "Sufficiently regular" means that all the integrals which we write in the following equations have a sense. For more details, see *Evequoz* (1980)

In the potential and kinetic plasma energies of (4.4) we replace the variables X, V, and Z as follows;

$$W_p=\frac{1}{2}\int_0^1\int_0^{2\pi}\frac{dr}{r}d\theta\left\{\frac{B_\theta^2}{r^2}\left|\frac{\partial X^{(1)}}{\partial\theta}+ikqX^{(2)}\right|^2\right.$$
$$+B_z^2\left|\frac{\partial X^{(3)}}{\partial r}+\frac{\partial V}{\partial\theta}\right|^2+B_\theta^2\left|\frac{\partial X^{(3)}}{\partial r}-ikqV^{(2)}-\frac{2}{r}X^{(2)}\right|^2$$
$$+\gamma p\left|\frac{\partial X^{(3)}}{\partial r}+\frac{\partial V}{\partial\theta}+\frac{B_\theta}{r}\left(\frac{\partial Z}{\partial\theta}+ikqZ^{(2)}\right)\right|^2$$
$$\left.-\frac{2B_\theta}{r^2}\frac{d}{dr}(rB_\theta)|X^{(2)}|^2\right\} \tag{4.24}$$

and

$$K=\frac{1}{2}\int_0^1\int_0^{2\pi}\varrho\frac{dr}{r}d\theta\{|X^{(2)}|^2+|rV^{(2)}+B_\theta Z^{(2)}|^2+B_z^2|Z^{(2)}|^2\} \ . \tag{4.25}$$

Recall that $q=rB_z/B_\theta$.

The variational formulation (4.3, 4) can be replaced by the following;

"Find real numbers ω^2 and non-trivial $\boldsymbol{u}=(X, X^{(1)}, X^{(2)}, X^{(3)}, V, V^{(2)}, Z, Z^{(2)})\in U$ satisfying $\delta\mathscr{L}=\delta(W_p-\omega^2K)=0$." (4.26)

The whole formulation includes eight functions altogether $X, X^{(1)}, X^{(2)}, X^{(3)}, V, V^{(2)}, Z$, and $Z^{(2)}$, and there are five integral conditions. Our aim now is to introduce a finite element expansion for all eight variables, then to eliminate $X^{(1)}, X^{(2)}, X^{(3)}$, by X, $V^{(2)}$ by V and $Z^{(2)}$ by Z through the integral relations (4.16–18, 20 and 22) when we choose the test function η in a finite-dimensional space. The only remaining variables will be X, V, and Z.

We introduce a finite-dimensional subspace U_h, the set of all piecewise polynomial functions $\boldsymbol{u}_h=(X_h, X_h^{(1)}, X_h^{(2)}, X_h^{(3)}, V_h, V_h^{(2)}, Z_h, Z_h^{(2)})$ with the same regularities and boundary conditions as given in (4.15–22), and with the integral conditions

$$\begin{aligned}
&\int_0^1\int_0^{2\pi}(X_h^{(1)}-X_h)\eta_h dr\,d\theta=0\\
&\int_0^1\int_0^{2\pi}(X_h^{(2)}-X_h)\eta_h dr\,d\theta=0\\
&\int_0^1\int_0^{2\pi}(X_h^{(3)}-X_h)\eta_h dr\,d\theta=0\\
&\int_0^1\int_0^{2\pi}(V_h^{(2)}-V_h)\eta_h dr\,d\theta=0\\
&\int_0^1\int_0^{2\pi}(Z_h^{(2)}-Z_h)\eta_h dr\,d\theta=0\ ,
\end{aligned} \tag{4.27}$$

for all η_h in the finite-dimensional subspace of $L^2(\Omega)$ of piecewise polynomial functions. The approximated problem is then just (4.26) with U replaced by U_h and $\boldsymbol{u}$ by $\boldsymbol{u}_h$.

4.2.2 Lowest-Order Finite Hybrid Elements

If we choose the lowest-order sufficiently regular elements, and divide Ω into $N_r * N_\theta$ cells (Fig. 4.2), then the eight unknowns X_h, $X_h^{(1)}$, $X_h^{(2)}$, $X_h^{(3)}$, V_h, $V_h^{(2)}$, Z_h, and $Z_h^{(2)}$ are represented by

$$
\begin{aligned}
X_h(r,\theta) &= \sum_{i=1}^{N_r-1} e_i(r) \sum_{j=0}^{N_\theta} X_{ij} e_j(\theta) \\
X_h^{(1)}(r,\theta) &= \sum_{i=1}^{N_r} c_{i-1/2}(r) \sum_{j=0}^{N_\theta} X_{i-1/2,j} e_j(\theta) \\
X_h^{(2)}(r,\theta) &= \sum_{i=1}^{N_r} c_{i-1/2}(r) \sum_{j=1}^{N_\theta} X_{i-1/2,j-1/2} c_{j-1/2}(\theta) \\
X_h^{(3)}(r,\theta) &= \sum_{i=0}^{N_r} e_i(r) \sum_{j=1}^{N_\theta} X_{i,j-1/2} c_{j-1/2}(\theta) \\
\begin{pmatrix} V_h(r,\theta) \\ Z_h(r,\theta) \end{pmatrix} &= \sum_{i=1}^{N_r} c_{i-1/2}(r) \sum_{j=0}^{N_\theta} \begin{pmatrix} V_{i-1/2,j} \\ Z_{i-1/2,j} \end{pmatrix} e_j(\theta) \\
\begin{pmatrix} V_h^{(2)}(r,\theta) \\ Z_h^{(2)}(r,\theta) \end{pmatrix} &= \sum_{i=1}^{N_r} c_{i-1/2}(r) \sum_{j=1}^{N_\theta} \begin{pmatrix} V_{i-1/2,j-1/2} \\ Z_{i-1/2,j-1/2} \end{pmatrix} c_{j-1/2}(\theta) \ .
\end{aligned}
\tag{4.28}
$$

The functions $c_{k-1/2}(x)$ and $e_k(x)$ are the piecewise constant elements defined in (1.39) and the hat functions defined in (1.18, 19, 37), respectively. The positions of the nodal points introduced in (4.28) are shown in Fig. 4.5.

Using

$$\eta_h(r,\theta) = c_{i-1/2}(r)\, c_{j-1/2}(\theta) \ , \tag{4.29}$$

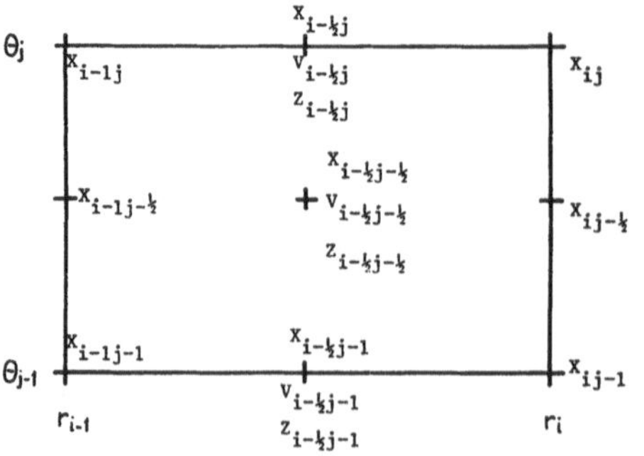

Fig. 4.5. Positions of the nodal points in the interval $r_{i-1} \le r \le r_i$, $\theta_{j-1} \le \theta \le \theta_j$

the integral conditions (4.27) with (4.28) lead to

$$
\begin{aligned}
X_{i-1/2,j} &= \frac{X_{i-1,j}+X_{ij}}{2} \\
X_{i-1/2,j-1/2} &= \frac{X_{i-1,j-1}+X_{i-1,j}+X_{i,j-1}+X_{ij}}{4} \\
X_{i,j-1/2} &= \frac{X_{i,j-1}+X_{ij}}{2} \\
V_{i-1/2,j-1/2} &= \frac{V_{i-1/2,j-1}+V_{i-1/2,j}}{2} \\
Z_{i-1/2,j-1/2} &= \frac{Z_{i-1/2,j-1}+Z_{i-1/2,j}}{2} \; .
\end{aligned}
\tag{4.30}
$$

The derivatives of X are very simple. They are

$$
\begin{aligned}
\frac{\partial X^{(1)}}{\partial\theta} &= \frac{X_{ij}+X_{i-1,j}-X_{i,j-1}-X_{i-1,j-1}}{2(\theta_j-\theta_{j-1})} \\
\frac{\partial X^{(3)}}{\partial r} &= \frac{X_{ij}+X_{i,j-1}-X_{i-1,j}-X_{i-1,j-1}}{2(r_i-r_{i-1})} \; .
\end{aligned}
\tag{4.31}
$$

We have to note that all quantities in (4.24 and 25) involving the unknowns are piecewise constant in a cell, and are equal to the value at the middle of the cell. All the geometric and equilibrium quantities are also taken to be constant in each mesh cell, and equal to the value at the middle of the cell. The integration of (4.24 and 25) can then be replaced by a sum over all cells, of the integrands taken at the center of each cell, times the size of the cell, i.e.,

$$
W_p = \tfrac{1}{2} \sum_{i=1}^{N_r} \sum_{j=1}^{N_\theta} (r_i - r_{i-1})(\theta_j - \theta_{j-1}) I(r_{i-1/2}, \theta_{j-1/2}) \; , \tag{4.32}
$$

where $I(r_{i-1/2}, \theta_{j-1/2})$ denotes the integrand at the mid point of the cell.

4.2.3 Application to the Test Cases

We compare the results of this method with those obtained in Chap. 3 and Sect. 4.1.2 and start by applying it to test case A for a homogeneous currentless plasma cylinder. Again, as with the previous "unpolluted" methods we find a spectrum with the right degeneracy $\omega^2 = k^2$.

Now we consider test case D which was analysed earlier in Chap. 4 using two dimensional conforming elements, the results of which were reproduced in Fig. 4.4. Applying our non-conforming two-dimensional finite hybrid elements,

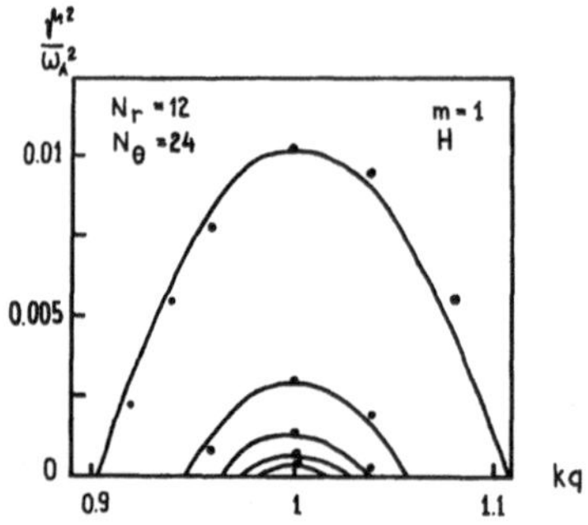

Fig. 4.6. The first five unstable modes of test case D (—). (···) are the eigenvalues obtained with *finite hybrid elements* using $N_r = 12$ and $N_\theta = 24$ intervals. (From *Gruber* 1978)

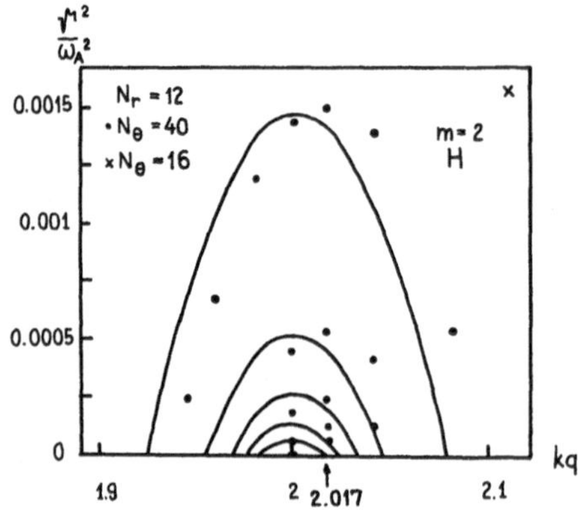

Fig. 4.7. The first five unstable modes of test case E (—). (···) are the eigenvalues obtained with finite hybrid elements using $N_r = 12$ and $N_\theta = 40$ intervals. Note that no unstable mode was obtained when conforming elements were taken. (From *Gruber* 1978)

we obtain the results shown in Fig. 4.6. We find that in the region $kq < 1$, the eigenvalues obtained lie below the "exact" curve, whereas they lie above it for $kq > 1$. The maximum of the curve $\Gamma^2(kq)$ is slightly shifted towards higher $|kq|$ values.

Instead of $kq = 1$, we find a maximum at $kq = 1{,}006$. With $N_r = 12$ and $N_\theta = 24$, the eleven unstable modes which should appear at $kq = 1$ are found at $kq = 1{,}006$. This means that the two-dimensional finite hybrid element method reproduces well the unstable part of the test case D, but it slightly modifies the physical quantity kq. We call this phenomenon a spectral shift.

This spectral shift becomes more striking when we consider the test case E for which, with the resolution taken, we were unable to find an unstable mode with two-dimensional conforming elements. However, with our finite hybrid elements, we are able to reproduce this spectrum quite accurately. In Fig. 4.7 we compare the 1D result (solid curve) with the 2D results (dots) obtained with our non-conforming approach. Using $N_r = 12$ and $N_\theta = 40$ intervals, the maximum of $\omega^2(kq)$ is obtained at $kq = 2{,}017$ instead of 2. At $kq = 2{,}017$, the eleven expected unstable modes can again be found.

4.2.4 Explanation of the Spectral Shift

The observed spectral shift depends on the number of poloidal intervals N_θ. This can be seen in Fig. 4.8. There we plot the difference between m and the kq

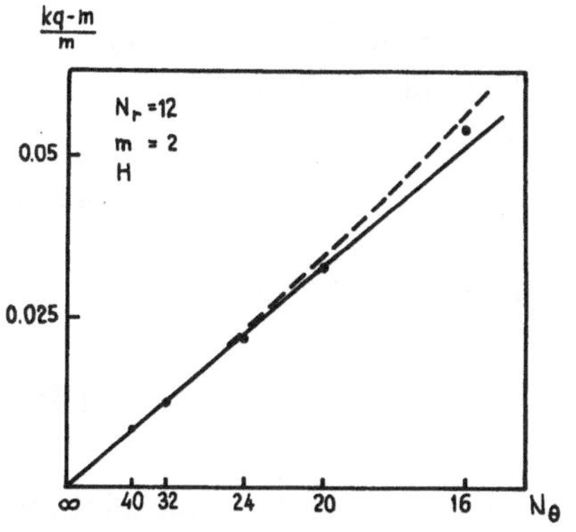

Fig. 4.8. The relative shift $(kq-m)/m$ for test case E approximated by finite hybrid elements. The straight solid line corresponds to quadratic convergence and the dotted line corresponds to the behaviour of the shift given by (4.42)

value at which the maximum growthrate squared occurs (for instance 0.017 for $N_\theta=40$) as a function of the number of poloidal intervals N_θ. The dotted straight line corresponds to quadratic convergence.

Where does this spectral shift come from? *Takizuka* et al. (1981a) give the answer. It is due to the approximation of the $\boldsymbol{B}\cdot\nabla$ operator (4.14) with finite hybrid elements. At $-kq=m$ this operator identically vanishes. The poloidal wave form is $\exp(im\theta)$. Let us now study what happens when we discretize (4.14) using finite hybrid elements. The two coupled equations fulfilling $\boldsymbol{B}\cdot\nabla X=0$ are (for r fixed)

$$\begin{aligned}&\frac{\partial X^{\mathrm{R}}}{\partial\theta}-kqX^{\mathrm{I}}=0\\&\frac{\partial X^{\mathrm{I}}}{\partial\theta}+kqX^{\mathrm{R}}=0\ ,\end{aligned}\tag{4.33}$$

where X^{R} and X^{I} are the real and imaginary parts of X, respectively. Eliminating X^{I} gives

$$\frac{\partial^2 X^{\mathrm{R}}}{\partial\theta^2}-k^2q^2X^{\mathrm{R}}=0\ .\tag{4.34}$$

X^{R} is restricted by the periodicity condition

$$X^{\mathrm{R}}(\theta=0)=X^{\mathrm{R}}(\theta=2\pi)\ .\tag{4.35}$$

Equations (4.34, 35) represent an eigenvalue problem with solutions

$$\begin{aligned}&k^2q^2=m^2\,,\qquad m=\text{integer}\\&X^{\mathrm{R}}(\theta)=X_0^{\mathrm{R}}\sin(m\theta)\ .\end{aligned}\tag{4.36}$$

We have taken the antisymmetric solution since in the MHD problem X^{R} is antisymmetric. The symmetric solution gives X^{I}.

The domain $0<\theta<2\pi$ is divided into N_θ equidistant intervals. Using finite hybrid elements the second derivative at $\theta=\theta_j$ is approximated by

$$\left(\frac{\partial^2 X^{\mathrm{R}}}{\partial\theta^2}\right)_h=\frac{X_{j-1}-2X_j+X_{j+1}}{\Delta\theta^2}\ , \tag{4.37}$$

where $\Delta\theta=2\pi/N_\theta$ is the interval size. The function itself becomes

$$(X^{\mathrm{R}})_h=\frac{X_{j-1}+2X_j+X_{j+1}}{4}\ . \tag{4.38}$$

We replace the nodal values in (4.37 and 38) by the values of the exact solution (4.36), leading to

$$\left(\frac{\partial^2 X^{\mathrm{R}}}{\partial\theta^2}\right)_h=X_0\frac{2\sin jm\Delta\theta[\cos m\Delta\theta-1]}{\Delta\theta^2} \tag{4.39}$$

and

$$(X^{\mathrm{R}})_h=X_0\frac{2\sin jm\Delta\theta[\cos m\Delta\theta+1]}{4}\ . \tag{4.40}$$

The approximated solution is equal to the exact one if the eigenvalue $-kq=m$ is modified to $\tilde{m}$ such that

$$\frac{\cos m\Delta\theta-1}{\Delta\theta^2}-\tilde{m}\frac{2\cos m\Delta\theta+1}{4}=0\ , \tag{4.41}$$

and

$$\tilde{m}=m\frac{\mathrm{tg}(m\Delta\theta/2)}{(m\Delta\theta/2)}\ . \tag{4.42}$$

The relative shift $(\tilde{m}-m)/m$ is

$$\frac{\tilde{m}-m}{m}=\frac{\mathrm{tg}(m\Delta\theta/2)}{(m\Delta\theta/2)}-1\ . \tag{4.43}$$

Expanding (4.43) for large N_θ or small argument in the tangens leads to the formula

$$\frac{\tilde{m}}{m}\cong\frac{1}{1-0.8225(2m/N_\theta)^2-0.1353(2m/N_\theta)^4}\ , \tag{4.44}$$

which corresponds to that reported in *Gruber* and *Troyon* (1977). This formula (4.44) has been applied to the case shown in Fig. 4.8 and plotted as a solid line. We see that it corresponds to the numerically determined shifts.

4.2.5 Convergence Properties

It has been shown by *Jaccard* (1980) and *Evequoz* (1980) that the finite hybrid element approach described leads to an unpolluted spectrum. Convergence is shown to be generally $O(h^2)$. If we correct for the spectral shift, superconvergence in $O(h^4)$ can sometimes be observed. Such a case is discussed by *Gruber* (1978). When taking away the shift, one often sees a convergence from "below", i.e., the discretization destabilizes the spectrum. This is in contrast to the conforming elements which always approach the lowest eigenvalue from "above". We have seen that this destabilization is a major advantage in the calculation of more general cases in toroidal and helical geometries, since instabilities can be found even with a few intervals. However, parts of the continuous spectrum, which is always stable, can become unstable due to the finite hybrid elements. Sometimes many radial intervals have to be taken to be able to predict if an observed instability is real or is part of the continuum. The convergence characteristics also depend on the profiles. In most of our calculations we observe destabilization of the continuum in a tokomak ($dq/d\psi > 0$), and stabilization of the spectrum for a spheromak ($dq/d\psi < 0$) (e.g., *Gautier* et al. 1981).

4.3 Discussion

In the numerical approximation of the two-dimensional MHD stability problem three difficulties arise:

One has to choose a finite element approach which lies in a function class satisfying $\nabla \cdot \xi = 0$, i.e., one has to prevent spectral pollution.

As a rule of thumb, the number of mesh cells should be comparable to the number of unknowns per vector component. This condition guarantees enough degrees of freedom to satisfy $\nabla \cdot \xi = 0$ everywhere. It eliminates triangularization (number of mesh cells ~ 2 times number of unknowns per component) as a viable method.

Singular ψ-surfaces on which the operator F vanishes correspond to surfaces at which singular and marginal solutions ($\omega^2 = 0$) occur. If we are not able to guarantee that F can reach zero, a strong stabilization effect can be observed for the unstable modes and, often, an unstable solution is stabilized up to a certain discretization. This precision problem arises with two-dimensional conforming quadrangular finite elements. As a consequence, we reject them and propose what we call "finite hybrid elements". However, interpretation of the convergence properties of this method has to be made carefully.

5. ERATO: Application to Toroidal Geometry

5.1 Static Equilibrium

5.1.1 Grad–Schlüter–Shafranov Equation

The calculation of static MHD equilibria in toroidal geometry is routinely used to answer different kinds of questions:

The engineer who builds a tokamak experiment has to run an equilibrium code to know where to put the magnetic field coils and to find out which currents have to be driven in order to produce a plasma with given characteristics.

The experimentalist who works with such a machine knows the positions of the field coils and the currents flowing in them. From the diagnostics, he obtains some information about the plasma. From this experimental data he should reconstitute relevant physical quantities in the plasma, such as mass density, temperature, and current density. This is not a simple undertaking since not much information is available from the interior of a hot plasma.

If stability studies have to be performed, one assumes that experimentalists have already found an equilibrium solution which fits the experimental data well. This solution enables one to know the plasma surface Γ_p as well as the pressure and current profiles in Ω_p. Since one has to know the equilibrium solution with high accuracy for a stability calculation, one recalculates a fixed boundary equilibrium starting from these given profiles.

In axisymmetric toroidal geometry such a fixed boundary magnetohydrostatic equilibrium can be solved in an (r, z, ϕ) coordinate system (Fig. 5.1). The symmetry axis is given by $r=0$. The equilibrium solution is homogeneous in ϕ. This means that for each cut with a surface through the symmetry axis we obtain the same equilibrium solution; the coordinate ϕ can be ignored.

Let us recall the three equilibrium equations (2.5) (the index 0 is omitted)

$$\nabla p = \boldsymbol{J} \times \boldsymbol{B} \ , \tag{5.1}$$

$$\nabla \times \boldsymbol{B} = \boldsymbol{J} \ , \tag{5.2}$$

$$\nabla \cdot \boldsymbol{B} = 0 \ , \tag{5.3}$$

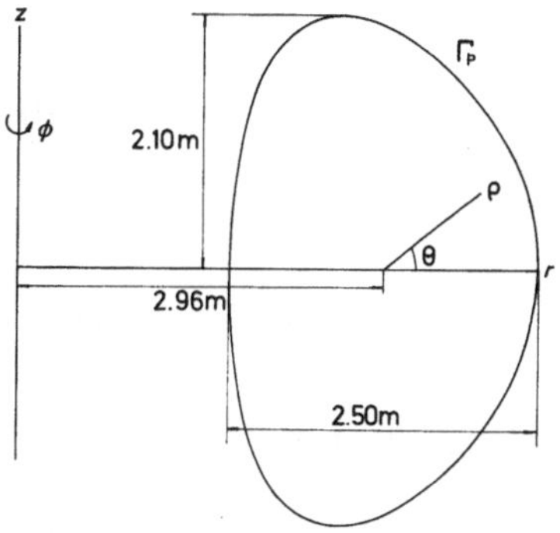

Fig. 5.1. The cylindrical coordinates in toroidal geometry. JET configuration

where $p=p(r,z)$, $\boldsymbol{J}=\boldsymbol{J}(r,z)$ and $\boldsymbol{B}=\boldsymbol{B}(r,z)$ are the plasma pressure, the current density and the magnetic field, respectively.

In axisymmetric geometry, a general solution for the magnetic field $\boldsymbol{B}$ which fulfils (5.3) is (*Greene* and *Johnson* 1962)

$$\boldsymbol{B}=T\nabla\phi+\nabla\phi\times\nabla\psi \;, \tag{5.4}$$

where ψ and T are the fluxes of the toroidal and the poloidal currents, respectively, and

$$\begin{aligned}\nabla\psi &\equiv \boldsymbol{e}_r\frac{\partial\psi}{\partial r}+\boldsymbol{e}_z\frac{\partial\psi}{\partial z}\\ \nabla\phi &\equiv \boldsymbol{e}_\phi\frac{1}{r} \;.\end{aligned} \tag{5.5}$$

Here, $\boldsymbol{e}_r$, $\boldsymbol{e}_z$, and $\boldsymbol{e}_\phi$ are the unit vectors in the r, z, and ϕ directions. The current density then is

$$\boldsymbol{J}=\nabla\times\boldsymbol{B}=-\nabla\phi\times\nabla T+r^2\nabla\cdot\left(\frac{\nabla\psi}{r^2}\right)\nabla\phi \tag{5.6}$$

and

$$\begin{aligned}\boldsymbol{J}\times\boldsymbol{B}=&-\frac{1}{r^2}T\nabla T-\left[\nabla\cdot\left(\frac{\nabla\psi}{r^2}\right)\right]\nabla\psi\\ &-[(\nabla\psi\times\nabla T)\cdot\nabla\phi]\nabla\phi \;.\end{aligned} \tag{5.7}$$

Since we impose axisymmetry, i.e.,

$$\boldsymbol{e}_\phi\cdot\boldsymbol{J}\times\boldsymbol{B}=0 \quad \text{or} \tag{5.8}$$

$$\nabla\psi\times\nabla T=0 \tag{5.9}$$

it follows that[1]

$$T = T(\psi(r,z)) \ . \tag{5.10}$$

The right-hand side of (5.1)

$$\boldsymbol{J} \times \boldsymbol{B} = \left[-\frac{1}{r^2} T \frac{dT}{d\psi} - \nabla \cdot \left(\frac{\nabla \psi}{r^2} \right) \right] \nabla \psi \tag{5.11}$$

has then only a component parallel to $\nabla\psi$. As a consequence,

$$p = p(\psi(r,z)) \, . \tag{5.12}$$

The partial differential equation obtained

$$\nabla \cdot \left(\frac{\nabla \psi}{r^2} \right) = -j/r \tag{5.13}$$

is called the Grad–Schlüter–Shafranov (GSS) equation; the toroidal current density

$$j \equiv \boldsymbol{e}_\phi \cdot \nabla \times \boldsymbol{B} = r \frac{dp}{d\psi} + \frac{1}{2r} \frac{dT^2}{d\psi} \tag{5.14}$$

contains the two arbitrary functions of ψ: $p(\psi(r,z))$ and $T^2(\psi(r,z))$. It vanishes at Γ_p in relevant physical situations. Choosing the fixed boundary case, i.e.,

$$\psi = 0 \quad \text{on } \Gamma_p \ , \tag{5.15}$$

and $\psi < 0$ in Ω_p, (5.13) can be satisfied by the trivial solution $\psi = 0$. To avoid this solution we have to introduce a normalization condition such as imposing the total toroidal current I defined by

$$I = \int_{\Omega_p} j \, dx \ , \tag{5.16}$$

where dx is the area element in Ω_p. In reality, we do not know explicitly the functions $p(\psi)$ and $T^2(\psi)$ which define j in (5.14). In practice, we impose $p^*(\psi)$ and $T^{*2}(\psi)$ such that

$$\begin{aligned} p(\psi) &= \lambda p^*(\psi) \\ T^2(\psi) &= \lambda T^{*2}(\psi) \ , \end{aligned} \tag{5.17}$$

where λ is a real parameter introduced to satisfy (5.16). We denote this by

$$j^* = r \frac{dp^*}{d\psi} + \frac{1}{2r} \frac{dT^{*2}}{d\psi} = \frac{1}{\lambda} j \ . \tag{5.18}$$

---

1 We admit here that $\nabla\psi = 0$ only at one point in the plasma domain

5.1.2 Weak Formulation

Problems (5.13, 16) can be written in variational form:

"Find ψ sufficiently regular in Ω_p satisfying (5.15) and a real value λ such that

$$\int_{\Omega_p} \frac{1}{r} \nabla\eta \cdot \nabla\psi \, dx = \lambda \int_{\Omega_p} \eta j^* dx \ , \tag{5.19}$$

for all η fulfilling the same conditions as ψ and where, for a given I

$$\lambda \int_{\Omega_p} j^* dx = I \ ." \tag{5.20}$$

Equations (5.19, 20) constitute a non-linear eigenvalue problem with a linear left-hand side.

In our computer code CLIO (*Semenzato* et al. 1984) it is solved in the following way:

A) As input quantities are prescribed:
i) The functional dependence of the plasma surface

$$\begin{aligned} r_p &= R_0 + \varrho_p(\theta)\cos\theta \\ z_p &= \varrho_p(\theta)\sin\theta \ , \end{aligned} \tag{5.21}$$

where $\varrho_p(\theta)$ is a function of θ with $0 \le \theta \le 2\pi$ (see Fig. 5.1).
ii) The total current I and the parameters p_i and t_i of the profiles

$$p(\psi) = \lambda \sum_{i=2}^{n} p_i \psi^i \ , \tag{5.22}$$

$$T^2(\psi) = T_{\text{lim}}^2 + \lambda \sum_{i=2}^{n} t_i \psi^i \ , \tag{5.23}$$

where T_{lim}^2 is the toroidal magnetic flux at Γ_p. In cases to be discussed later in this chapter we shall take $n = 3$.

B) The non-linearity appearing in the right-hand side of (5.19) is treated by a Picard method. One starts from an initial guess ψ_0. If (ψ_k, λ_k) is the solution pair after k Picard steps, the updated solution pair $(\psi_{k+1}, \lambda_{k+1})$ is obtained by solving for all η with $\eta = 0$ on Γ_p

$$\int_{\Omega_p} \frac{1}{r} \nabla\eta \cdot \nabla\psi_{k+1} dx = \lambda_k \int_{\Omega_p} \eta j_k^* dx \quad \text{and} \tag{5.24}$$

$$\lambda_{k+1} \int_{\Omega_p} j_{k+1}^* dx = I \ , \tag{5.25}$$

where

$$j^*_{k+1}=r\sum_{i=2}^{n} ip_i\psi_{k+1}^{i-1}+\frac{1}{2r}\sum_{i=2}^{n} it_i\psi_{k+1}^{i-1} \ . \tag{5.26}$$

C) To fit best the functional dependence $\varrho_p(\theta)$ of the plasma surface (5.21) we introduce the independent non-orthogonal coordinates σ and θ which are related to r and z by

$$\begin{aligned} r&=R_0+\sigma\varrho_p(\theta)\cos\theta \\ z&=\sigma\varrho_p(\theta)\sin\theta \ . \end{aligned} \tag{5.27}$$

The coordinate σ varies from 0 at the axis $r=R_0$ and $z=0$ to $\sigma=1$ at the plasma surface Γ_p.

D) In (σ,θ) coordinates, the linear problem (5.24) is written:

$$\begin{aligned} &\int_0^1\int_0^{2\pi}\frac{\sigma}{r}\,d\sigma\,d\theta\left\{\frac{\partial\psi_{k+1}}{\partial\sigma}\frac{\partial\eta}{\partial\sigma}+\left(\frac{1}{\sigma}\frac{\partial\psi_{k+1}}{\partial\theta}-\frac{d\varrho_p/d\theta}{\varrho_p}\frac{\partial\psi_{k+1}}{\partial\sigma}\right)\right.\\ &\qquad\left.\cdot\left(\frac{1}{\sigma}\frac{\partial\eta}{\partial\theta}-\frac{d\varrho_p/d\theta}{\varrho_p}\frac{\partial\eta}{\partial\sigma}\right)\right\}\\ &\quad=\lambda_k\int_0^1\int_0^{2\pi}\eta j^*_k\varrho_p^2\sigma\,d\sigma\,d\theta \ . \end{aligned} \tag{5.28}$$

The domain $\Omega_p=\{0\leqq\sigma\leqq1, 0\leqq\theta\leqq2\pi\}$ is subdivided into $N_\sigma\times N_\theta$ quadrangles. The dependent variable ψ_{k+1} in (5.28) is approximated by isoparametric linear finite elements (for details, see *Semenzato* et al. 1984).

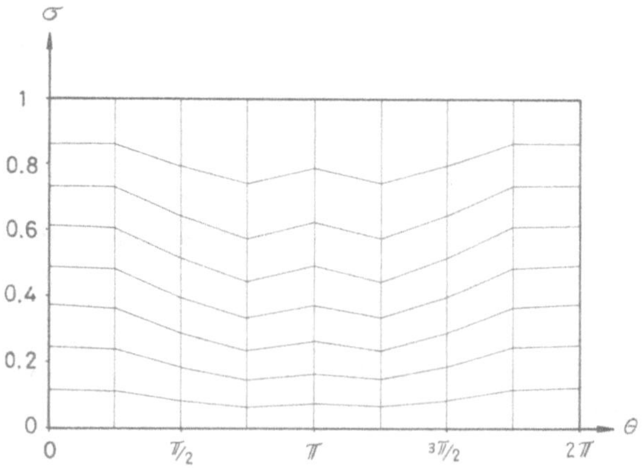

Fig. 5.2. Subdivision of the (σ,θ) plane in $N_\sigma * N_\theta$ quadrangles ($N_\sigma=8$, $N_\theta=8$ in this case). $\sigma=1$ is the plasma surface Γ_p and $\sigma=0$ is the magnetic axis

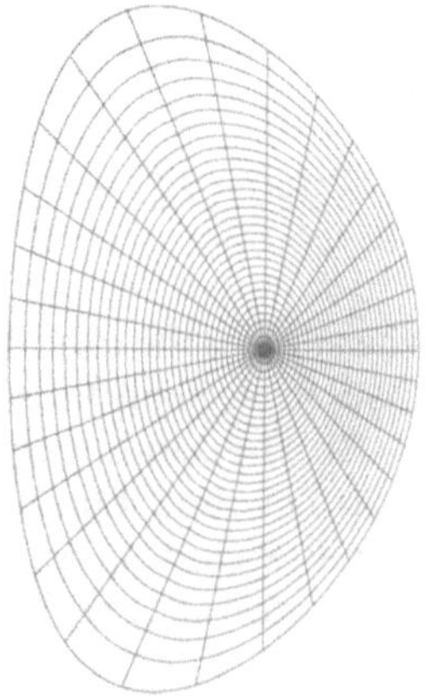

Fig. 5.3. Flux surface plot in the (r, z) plane of the equilibrium solution, calculated in the (σ,θ) plane

E) During the iteration process on the non-linearity the grid points are adjusted such that all the nodal points fall on initially prescribed values of ψ. As an example the final (σ, θ) grid of a JET equilibrium solution is shown in Fig. 5.2. The corresponding representation in the (r, z) plane is given in Fig. 5.3. The value of ψ at the magnetic axis is $\psi = \psi_p$. Note that ψ_p is the total poloidal flux in Ω_p.

5.2 Mapping of (σ, θ) into (ψ, χ) Coordinates in Ω_p

For the calculation of MHD stability we have to have some information about the equilibrium solution. As seen previously, the equilibrium code delivers a solution $\psi(\sigma, \theta)$ in which the σ coordinate is arranged to match $\psi = \text{const}$ surfaces. These surfaces are chosen as one of the two independent variables for the stability calculation. The other independent variable which we call χ is chosen in such a way that the Jacobian J is

$$J = \frac{r}{|\nabla \psi \times \nabla \chi|} = \frac{q(\psi)}{T(\psi)} r^2 \, , \tag{5.29}$$

where $q(\psi)$ is a function of ψ, and $T(\psi)$ is the arbitrary function introduced in (5.4). On each $\psi = \text{const}$ surface the arc length dl is related to the change $d\chi$ of χ by (Fig. 5.4)

$$d\chi = \nabla \chi \cdot d\mathbf{l} = \nabla \chi \cdot \left[\frac{r \, dl}{|\nabla \psi|} (\nabla \psi \times \nabla \phi) \right]$$

$$= \frac{r \, dl}{J |\nabla \psi|} \, . \tag{5.30}$$

We demand that χ be a 2π periodic function with respect to θ, i.e.,

$$\oint \frac{r}{J |\nabla \psi|} \, dl = 2\pi \, , \tag{5.31}$$

$$d\underline{l} = \frac{\nabla \Psi \times \nabla \varphi}{|\nabla \Psi|} \, r d\ell$$

$$dX = d\underline{l} \times \nabla \chi = \frac{r}{J |\nabla \Psi|} \, d\ell$$

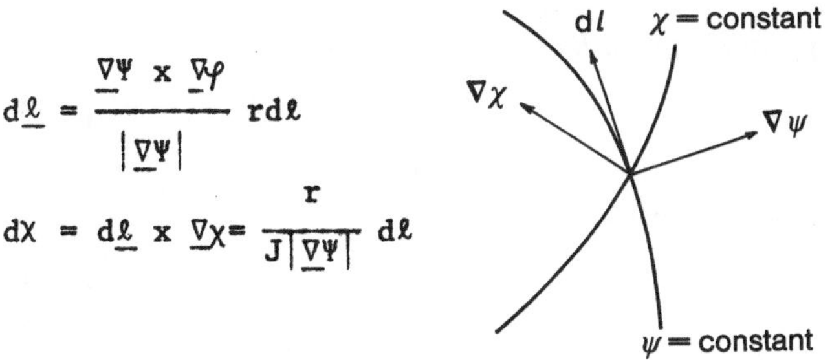

Fig. 5.4. Relation between the χ-coordinate and the arc length on a $\psi = \text{const}$ surface

where the integral path is along a cut of the meredian plane with a $\psi = \text{const}$ surface. The integral (5.31) thus defines the quantity $q(\psi)$ which is called the safety factor and is introduced in (5.29):

$$q(\psi) = \frac{T(\psi)}{2\pi} \oint \frac{dl}{r|\nabla\psi|} \ . \tag{5.32}$$

It is convenient to introduce cylindrical coordinates (ϱ, θ) centered at the magnetic axis R_0. On a $\psi = \text{const}$ surface

$$dl = \frac{\varrho|\nabla\psi|}{\partial\psi/\partial\varrho} \, d\theta \ . \tag{5.33}$$

The coordinate χ expressed in θ on a $\psi = \text{const}$ surface is then

$$\chi(\theta) = \int_0^\theta \frac{\varrho \, d\theta}{J(\partial\psi/\partial\varrho)} \ . \tag{5.34}$$

5.3 Variational Formulation of the Potential and Kinetic Energies

As we already know from Chap. 2 the MHD stability problem can be written

$$\delta\mathcal{L} = \delta(W_p + W_v - \omega^2 K) = 0 \, , \tag{5.35}$$

where W_p is the potential energy in Ω_p given by[2]

$$\begin{aligned} W_p = \frac{1}{2} \int_{\Omega_p} J \, d\psi \, d\chi \{ & [\nabla \times (\boldsymbol{\xi} \times \boldsymbol{B}) \\ & + (\boldsymbol{n} \cdot \boldsymbol{\xi})(\boldsymbol{J} \times \boldsymbol{n})]^2 \\ & + \gamma p(\nabla \cdot \boldsymbol{\xi})^2 - 2(\boldsymbol{n} \cdot \boldsymbol{\xi})^2 (\boldsymbol{J} \times \boldsymbol{n})(\boldsymbol{B} \cdot \nabla)\boldsymbol{n} \} \ , \end{aligned} \tag{5.36}$$

W_v is the potential energy in Ω_v given by

$$W_v = \frac{1}{2} \int_{\Omega_v} J \, d\psi \, d\chi (\nabla \times \boldsymbol{A})^2 \tag{5.37}$$

and $-\omega^2 K$ is the kinetic energy with

$$K = \frac{1}{2} \int_{\Omega_p} J \, d\psi \, d\chi \, \varrho \xi^2 \ . \tag{5.38}$$

---

2 The domains Ω_p and Ω_v are shown in Fig. 2.1

In (5.37), $\boldsymbol{A}$ is the vector potential corresponding to the perturbed magnetic field, i.e., $\nabla\times\boldsymbol{A}=\delta\boldsymbol{B}$. The boundary condition at the conducting wall Γ_{v} is

$$\boldsymbol{n}\times\boldsymbol{A}=0 \quad \text{on} \quad \Gamma_{\mathrm{v}} \ . \tag{5.39}$$

The continuity condition at the plasma-vacuum interface Γ_{p} is

$$\boldsymbol{n}\times\boldsymbol{A}=-(\boldsymbol{n}\cdot\boldsymbol{\xi})\boldsymbol{B} \quad \text{on} \quad \Gamma_{\mathrm{p}} \ . \tag{5.40}$$

We refer to the *free boundary* case if $\Gamma_{\mathrm{v}}\neq\Gamma_{\mathrm{p}}$ and to the *fixed boundary* case if $\Gamma_{\mathrm{v}}=\Gamma_{\mathrm{p}}$, in which case,

$$\boldsymbol{n}\cdot\boldsymbol{\xi}=0 \quad \text{on} \quad \Gamma_{\mathrm{p}}=\Gamma_{\mathrm{v}} \ . \tag{5.41}$$

A convenient representation of the displacement vector in terms of the three components X, V, and Z is

$$\boldsymbol{\xi}=Xr^2(\nabla\chi\times\nabla\phi)+Vr^2(\nabla\phi\times\nabla\psi)+Zr^2\boldsymbol{B} \ . \tag{5.42}$$

With this choice of components

$$\begin{aligned}
&\boldsymbol{\xi}\times\boldsymbol{B}=VT\nabla\psi-XT\nabla\chi+X\frac{r^2}{J}\nabla\phi \\
&\nabla\psi\cdot\boldsymbol{\xi}=X\frac{r^2}{J} \ .
\end{aligned} \tag{5.43}$$

Note that the component Z parallel to $\boldsymbol{B}$ only appears in the $\nabla\cdot\boldsymbol{\xi}$ term. The four terms in W_{p} are then

$$\begin{aligned}
&\nabla\times(\boldsymbol{\xi}\times\boldsymbol{B}) \\
&\quad=-\nabla\psi\times\nabla(VT)+\nabla\chi\times\nabla(XT)-\nabla\phi\times\nabla\left(X\frac{r^2}{J}\right) , \\
&(\boldsymbol{n}\cdot\boldsymbol{\xi})(\boldsymbol{J}\times\boldsymbol{n}) \\
&\quad=\frac{Xr^2}{J|\nabla\psi|^2}\left[(\nabla\psi\cdot\nabla T)\nabla\phi-jr\nabla\phi\times\nabla\psi\right] , \\
&\nabla\cdot\boldsymbol{\xi}=(\nabla\chi\times\nabla\phi)\cdot\nabla(Xr^2) \\
&\qquad+(\nabla\phi\times\nabla\psi)\cdot\nabla(Vr^2)+(\boldsymbol{B}\cdot\nabla)(Zr^2) , \\
&2(\boldsymbol{J}\times\boldsymbol{n})\cdot(\boldsymbol{B}\cdot\nabla)\boldsymbol{n} \\
&\quad=(\boldsymbol{J}\times\boldsymbol{n})\cdot[\nabla\times(\boldsymbol{n}\times\boldsymbol{B})+\boldsymbol{B}(\nabla\cdot\boldsymbol{n})+\boldsymbol{J}\times\boldsymbol{n}] \ .
\end{aligned} \tag{5.44}$$

To calculate the squares in W_{p} and K, we project the terms onto an orthogonal coordinate system with unit vectors $\nabla\psi/|\nabla\psi|$, $\nabla\phi/|\nabla\phi|$ and $\nabla\phi\times\nabla\psi/(|\nabla\phi|\,|\nabla\psi|)$.

The projection of the first square in W_p is then

$$\begin{aligned}\nabla\times(\xi\times B)+(n\cdot\xi)(J\times n)\\ =A_1\frac{\nabla\psi}{|\nabla\psi|}+A_2\frac{\nabla\phi}{|\nabla\phi|}+A_3\frac{\nabla\phi\times\nabla\psi}{|\nabla\phi|\,|\nabla\psi|}\,,\end{aligned}\tag{5.45}$$

where

$$\begin{aligned}A_1&=\frac{1}{|\nabla\psi|}(B\cdot\nabla)(\xi\cdot\nabla\psi)\\ A_2&=-\frac{T}{|\nabla\phi|}[(\nabla\chi\times\nabla\phi)\cdot\nabla X\\ &\quad+(\nabla\phi\times\nabla\psi)\cdot\nabla V]\\ A_3&=\frac{|\nabla\psi|}{|\nabla\phi|}(\nabla\phi\cdot\nabla(VT))\\ &\quad-\frac{\nabla\chi\cdot\nabla\psi}{|\nabla\psi|^2}\frac{|\nabla\psi|}{|\nabla\phi|}(\nabla\phi\cdot\nabla(XT))\\ &\quad-\frac{|\nabla\phi|}{|\nabla\psi|}\left(\nabla\psi\cdot\nabla\left(\frac{Xr^2}{J}\right)\right)-\frac{jXr^2}{J|\nabla\psi|}\,.\end{aligned}\tag{5.46}$$

In axisymmetric geometry we demand that the equilibrium solution be homogeneous in the toroidal direction. The solution to the stability problem can then be Fourier-analyzed

$$\xi(\psi,\chi,\phi)=\sum_{n=-\infty}^{+\infty}\xi_n(\psi,\chi)\,e^{in\phi}\,.\tag{5.47}$$

One can show that different toroidal mode numbers n decouple and the stability problem can be solved separately for each individual n number. In practice, we often look for cavity modes with very low n numbers ($n=0$, 1, or perhaps 2) and check the stability of very high n number modes by a ballooning mode criterion. Intermediate n-number modes ($n=10, 20, \dots$) are sometimes calculated to understand the change in mode structure when going from low-n to high-n modes. The expressions in (5.46) are

$$\begin{aligned}\beta_\chi&=(\nabla\chi\cdot\nabla\psi)/|\nabla\psi|^2\\ \nabla&=\nabla\psi\frac{\partial}{\partial\psi}+\nabla\chi\frac{\partial}{\partial\chi}+in\nabla\phi\\ \nabla\phi\cdot\nabla&=in/r^2\\ \nabla\psi\cdot\nabla&=|\nabla\psi|^2\left(\frac{\partial}{\partial\psi}+\beta_\chi\frac{\partial}{\partial\chi}\right)\end{aligned}\tag{5.48}$$

$$J(\nabla\chi \times \nabla\phi) \cdot \nabla = \frac{\partial}{\partial\psi}$$

$$J(\nabla\phi \times \nabla\psi) \cdot \nabla = \frac{\partial}{\partial\chi}$$

$$\boldsymbol{B} \cdot \nabla = \frac{1}{J}\left(\frac{\partial}{\partial\chi} + in\frac{JT}{r^2}\right) \equiv \frac{1}{J}F \ .$$

We introduce the new independent variable

$$s = \sqrt{\frac{\psi_p + \psi}{\psi_p}} , \tag{5.49}$$

and the new dependent variables

$$\begin{aligned} \hat{X} &= X_n \\ \hat{V} &= 2s\psi_p V_n \\ \hat{Z} &= 2s\psi_p Z_n \ , \end{aligned} \tag{5.50}$$

and the quantities[3]

$$\begin{aligned} \hat{\beta}_\chi &= 2s\psi_p\beta_\chi = \left(\frac{\partial\chi}{\partial s}\right)_n \\ H &= \frac{J}{r^2}\left(\frac{\partial r^2/J}{\partial s}\right)_n + 2s\psi_p\frac{jr}{|\nabla\psi|^2} \ . \end{aligned} \tag{5.51}$$

Then the potential energy for a given n number becomes

$$\begin{aligned} W_p = \frac{1}{2}\int_0^1\int_0^{2\pi} ds\, d\chi\{&a_1|I_1|^2 + a_2|I_2|^2 \\ &+ a_3|I_3|^2 + a_4|I_4|^2 - a_5|I_5|^2\} \ , \end{aligned} \tag{5.52}$$

where

$$\begin{aligned} I_1 &= F[\hat{X}] = \frac{\partial\hat{X}}{\partial\chi} + in\frac{JT}{r^2}\hat{X} \\ I_2 &= D = \frac{\partial\hat{X}}{\partial s} + \frac{\partial\hat{V}}{\partial\chi} \\ I_3 &= D + H\hat{X} + \hat{\beta}_\chi F[\hat{X}] - F[\hat{V}] \\ I_4 &= D + \hat{X}\frac{\partial\ln r^2}{\partial s} + \hat{V}\frac{\partial\ln r^2}{\partial\chi} + \frac{1}{r^2}F[\hat{Z}r^2] \\ I_5 &= \hat{X} \end{aligned} \tag{5.53}$$

3 The lower index $\boldsymbol{n}$ indicate derivatives in normal directions

and

$$a_1=\frac{2s\psi_p r^4}{J^3|\nabla\psi|^2}$$

$$a_2=\frac{r^2}{2s\psi_p J}T^2$$

$$a_3=\frac{r^2}{2s\psi_p J}|\nabla\psi|^2 \tag{5.54}$$

$$a_4=\frac{r^2}{2s\psi_p J}r^2\gamma p$$

$$a_5=\frac{4s\psi_p r^4}{J}\left[-\frac{dp}{d\psi}\left(\frac{\partial\ln r}{\partial\psi}\right)_n + \frac{j^2}{|\nabla\psi|^2}+\frac{j}{2r}\left(\frac{\partial\ln|\nabla\psi|^2}{\partial\psi}\right)_n\right].$$

The kinetic energy K then becomes

$$K=\frac{1}{2}\int_0^1\int_0^{2\pi} ds\,d\chi\{b_1|I_6|^2+b_2|I_7|^2+b_3|I_8|^2\} \tag{5.55}$$

with

$$I_6=\hat{X}$$

$$I_7=-\beta_\chi\hat{X}+\hat{V}+\hat{Z} \tag{5.56}$$

$$I_8=\hat{Z} \quad \text{and,}$$

$$b_1=\frac{2s\psi_p\varrho}{J|\nabla\psi|^2}$$

$$b_2=\frac{Jr^2|\nabla\psi|^2\varrho}{2s\psi_p} \tag{5.57}$$

$$b_3=\frac{Jr^2T^2\varrho}{2s\psi_p}.$$

5.4 Variational Formulation of the Vacuum Energy

If Maxwell's equations have to be solved in a region bounded by given surfaces it is convenient to apply a Green's function technique. Such an approach has been chosen in the published version of the ideal linear MHD stability code

ERATO for a toroidal geometry (*Gruber* et al. 1981a). This Green's function technique has the advantage that a vacuum region extending to infinity can be treated easily. However, it suffers from major numerical disadvantages: the discretized Green's function is not in general symmetric; the integrals contain singularities which have to be extracted analytically; the convergence properties are different from those for the plasma contribution; the axisymmetric case $n=0$ has to be treated separately; coding is relatively complicated; and the computing time is large.

To overcome these problems, we propose to consider the vacuum as a shearless pseudoplasma with neither currents nor pressure (see *Gruber* et al. 1981b). We first introduce a coordinate system $(\tilde{\psi}(r,z), \tilde{\chi}(r,z))$ in the vacuum region Ω_v, similar to that used in Ω_p, which satisfies

$$\begin{aligned} &\tilde{\psi}=0 && \text{on} \quad \Gamma_p \\ &\tilde{\chi}=\chi_p && \text{on} \quad \Gamma_p \\ &\tilde{\psi}=\psi_v-\psi_p && \text{on} \quad \Gamma_v \,. \end{aligned} \tag{5.58}$$

To each point (r,z) there corresponds a unique point $(\tilde{\psi}, \tilde{\chi})$, and vice versa. We choose $\tilde{\psi}(r,z)$ and $\tilde{\chi}(r,z)$ in the following way (Fig. 5.5): Let us fix a point R_c (in general $R_c \neq R_0$) on the axis $z=0$ inside the plasma region Ω_p. Starting from this point, we draw straight lines to the poloidal mesh points χ_j on Γ_p. These rays are continued to Γ_v. Such a straight line from Γ_p to Γ_v is a $\tilde{\chi}=\text{const}$ line with the value identical to the χ value at Γ_p. This guarantees continuity of the coordinate χ through Γ_p. The coordinate $\tilde{\psi}$ is an increasing function from $\tilde{\psi}=0$ on Γ_p to $\tilde{\psi}=\psi_v-\psi_p$ on Γ_v on each of these $\tilde{\chi}=\text{const}$ rays. To guarantee this, we introduce $\varrho(r,z)$, or $\varrho(\tilde{\psi}, \tilde{\chi})$, which measures the distance of a point from R_c, and the two functions $\varrho_p(r,z)$ and $\varrho_v(r,z)$, or $\varrho_p(\tilde{\chi})$ and $\varrho_v(\tilde{\chi})$, which define Γ_p and Γ_v,

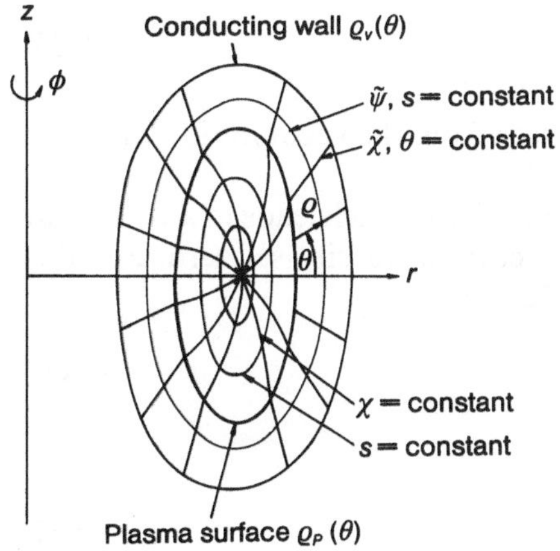

Fig. 5.5. The local coordinates in toroidal geometry

respectively. We can then define

$$\tilde{\psi}=\frac{\varrho(\tilde{\chi})-\varrho_p(\tilde{\chi})}{\varrho_v(\tilde{\chi})-\varrho_p(\tilde{\chi})}(\psi_v-\psi_p) \ . \tag{5.59}$$

We have chosen $\psi_v=4\psi_p$ such that our "radial" coordinate

$$s\equiv\sqrt{\frac{\tilde{\psi}+\psi_p}{\psi_p}} \tag{5.60}$$

varies between $1\leq s\leq 2$ in Ω_v. We remind the reader that $0\leq s\leq 1$ in Ω_p. In Ω_v, the Jacobian defined by

$$r\,dr\,dz=\tilde{J}\,d\tilde{\psi}\,d\tilde{\chi}=2s\psi_p\tilde{J}\,ds\,d\tilde{\chi} \tag{5.61}$$

is

$$\tilde{J}=\frac{\varrho r(\partial\psi/\partial\varrho)_p J_p}{\varrho_p r_p(\partial\tilde{\psi}/\partial\varrho)} \ . \tag{5.62}$$

Here the index p means evaluation of the quantity at the plasma side of Γ_p, i.e., at the point (ψ_p,χ_p). The derivatives are at constant $\tilde{\chi}$.

In this coordinate system we have to compute the vacuum energy

$$W_v=\frac{1}{2}\int_1^2\int_0^{2\pi}(\nabla\times A)^2 2s\psi_p\tilde{J}\,ds\,d\tilde{\chi} \ , \tag{5.63}$$

where A has to satisfy

$$\boldsymbol{n}\times\boldsymbol{A}=0 \quad \text{on} \quad \Gamma_v \ , \tag{5.64}$$

and

$$\boldsymbol{n}\times\boldsymbol{A}=(\boldsymbol{n}\cdot\boldsymbol{\xi})\boldsymbol{B} \quad \text{on} \quad \Gamma_p \ . \tag{5.65}$$

In this last expression the left-hand side is seen from the vacuum side while the right-hand side is seen from the plasma side.

Our goal now is to find a representation of $\nabla\times A$ which leads to a W_v similar to that of the plasma with a pressure and current density set zero. For this purpose we introduce a pseudomagnetic field $\tilde{\boldsymbol{B}}$

$$\tilde{\boldsymbol{B}}(\tilde{\psi},\tilde{\chi})=\tilde{T}(\tilde{\psi},\tilde{\chi})\nabla\phi+\nabla\phi\times\nabla\tilde{\psi} \tag{5.66}$$

which satisfies

$$\nabla\cdot\tilde{\boldsymbol{B}}=0 \ . \tag{5.67}$$

We demand that the operator $\tilde{\boldsymbol{B}}\cdot\nabla$ be invertible. This pseudofield is not the vacuum magnetic field since

$$\nabla\times\tilde{\boldsymbol{B}}\neq 0\ . \tag{5.68}$$

Since we only have to know $\nabla\times\boldsymbol{A}$ we can add or subtract a gradient[4] to $\boldsymbol{A}$ and introduce a pseudo-potential $\tilde{\boldsymbol{A}}$ by

$$\tilde{\boldsymbol{A}}=\boldsymbol{A}-\nabla\Phi\ . \tag{5.69}$$

Replacing $\boldsymbol{A}$ by $\tilde{\boldsymbol{A}}$ in (5.63) leads to the same vacuum energy. We can now choose Φ such that $\tilde{\boldsymbol{A}}$ has no $\tilde{\boldsymbol{B}}$ component, i.e.,

$$\tilde{\boldsymbol{B}}\cdot\tilde{\boldsymbol{A}}=\tilde{\boldsymbol{B}}\cdot\boldsymbol{A}-\tilde{\boldsymbol{B}}\cdot\nabla\Phi=0\ . \tag{5.70}$$

From (5.70) we could compute Φ since we demanded earlier that $\tilde{\boldsymbol{B}}\cdot\nabla$ be invertible. Since $\tilde{\boldsymbol{A}}$ now has no $\tilde{\boldsymbol{B}}$ component we can write

$$\tilde{\boldsymbol{A}}=\tilde{\xi}\times\tilde{\boldsymbol{B}} \tag{5.71}$$

and W_v becomes

$$W_v=\frac{1}{2}\int_1^2\int_0^{2\pi}[\nabla\times(\tilde{\xi}\times\tilde{\boldsymbol{B}})]^2 2s\psi_p\tilde{J}\,ds\,d\tilde{\chi}\ . \tag{5.72}$$

As a pseudodisplacement in Ω_v we choose

$$\begin{aligned}\tilde{\xi}(\tilde{\psi},\tilde{\chi})=&\frac{T_p}{q_p}\tilde{X}(\tilde{\psi},\tilde{\chi})\frac{\nabla\tilde{\psi}}{|\nabla\tilde{\psi}|}\\&+r\tilde{V}(\tilde{\psi},\tilde{\chi})\frac{\nabla\phi}{|\nabla\phi|}\ ,\end{aligned} \tag{5.73}$$

where $\tilde{X}$ denotes the component of ξ normal to $\psi=$const surfaces, $\tilde{V}$ is the toroidal component, and T_p and q_p are the toroidal magnetic flux and the safety factor in Ω_p at Γ_p. At the conducting wall Γ_v, the condition (5.64) is simply

$$\tilde{X}(\tilde{\psi}=\psi_v-\psi_p,\tilde{\chi})=0\quad\text{at}\quad\Gamma_v\ . \tag{5.74}$$

At the plasma-vacuum interface Γ_p where $\psi=0$ and $\chi=\chi_p$ the condition (5.65) becomes

$$\begin{aligned}&-\frac{T_p}{q_p|\nabla\psi|}\tilde{X}\left[\tilde{T}_p\frac{|\nabla\psi|}{|\nabla\tilde{\psi}|}\nabla\phi+\nabla\phi\times\nabla\psi\right]\\&=-\frac{T_p}{q_p|\nabla\psi|}X[T_p\nabla\phi+\nabla\phi\times\nabla\psi]\ .\end{aligned} \tag{5.75}$$

---

4 The gauge does not have to be known since we are not interested in knowing the scalar potential

The $\nabla\phi \times \nabla\psi$ component gives

$$\tilde{X}(\psi=0,\chi)=X(\psi=0,\chi) \quad \text{at} \quad \Gamma_{\rm p} , \tag{5.76}$$

and the $\nabla\phi$ component fixes

$$\tilde{T}_{\rm p}(\psi=0,\chi)=\frac{|\nabla\tilde{\psi}|}{|\nabla\psi|}T_{\rm p}=\frac{(\partial\tilde{\psi}/\partial\varrho)}{(\partial\psi/\partial\varrho)}T_{\rm p} \quad \text{at} \quad \Gamma_{\rm p} , \tag{5.77}$$

where the partial derivatives are evaluated at constant χ and $\tilde{\chi}$. To guarantee that $\tilde{\boldsymbol{B}}\cdot\nabla$ is invertible in $\Omega_{\rm v}$ the function $\tilde{T}$ has to be

$$\tilde{T}(\tilde{\psi},\tilde{\chi})=\frac{r^2 q_{\rm p}}{\tilde{J}} . \tag{5.78}$$

Substitution of all these expressions into $W_{\rm v}$ (5.72) leads to

$$W_{\rm v}=\frac{1}{2}\int_1^2\int_0^{2\pi} ds\, d\tilde{\chi}\left(\frac{T_{\rm p}}{\tilde{T}}\right)^2 \cdot(\tilde{a}_1|\tilde{I}_1|^2+\tilde{a}_2|\tilde{I}_2|^2+\tilde{a}_3|\tilde{I}_3|^2) , \tag{5.79}$$

where the constants $\tilde{a}_1$, $\tilde{a}_2$, and $\tilde{a}_3$ and the expressions $\tilde{I}_1$, $\tilde{I}_2$, and $\tilde{I}_3$ have the same forms as those in the plasma region, (5.52–54), with the only exception that $H=0$.

In the vacuum region one only has to vary with respect to two variables, $\tilde{X}$ and $\tilde{V}$. For the special case where $n=0$, the variable $\tilde{V}$ only appears in the second term of expression (5.79). Since there is no boundary condition on $\tilde{V}$, the variation with respect to $\tilde{V}$ can be performed. It leads to

$$\tilde{I}_2=\frac{\partial\tilde{X}}{\partial s}+\frac{\partial\tilde{V}}{\partial\tilde{\chi}}=0 \quad \text{for} \quad n=0 . \tag{5.80}$$

Expression (5.79) then becomes

$$W_{\rm v}=\frac{1}{2}\int_1^2\int_0^{2\pi} ds\, d\tilde{\chi}\left(\frac{T_{\rm p}}{\tilde{T}}\right)^2 \cdot\left\{\tilde{a}_1\left|\frac{\partial\tilde{X}}{\partial\tilde{\chi}}\right|^2+\tilde{a}_3\left|\frac{\partial\tilde{X}}{ds}+\tilde{\beta}_\chi\frac{\partial\tilde{X}}{\partial\tilde{\chi}}\right|^2\right\} \quad \text{for} \quad n=0 , \tag{5.81}$$

where $\tilde{\beta}_\chi=\left(\frac{\partial\tilde{\chi}}{\partial s}\right)_n$.

5.5 Finite Hybrid Elements

In toroidal geometry there exist localized solutions in Ω_p on integer nq $(=m)$ surfaces with $\omega^2=0$ and $|\hat{X}|\ll|\hat{V}|$ which satisfy

$$F[\hat{V}]=\frac{\partial \hat{V}}{\partial \chi}+inq\hat{V}=0$$

$$D=\frac{\partial \hat{X}}{\partial s}+\frac{\partial \hat{V}}{\partial \chi}=0 \tag{5.82}$$

$$\nabla\cdot\xi=\hat{V}\frac{\partial \ln r^2}{\partial \chi}+\frac{1}{r^2}F[r^2\hat{Z}]=0\ .$$

Such marginal solutions are part of the continuous spectrum. The components $\hat{X}$ and $\hat{V}$ have an angular variation of the form $(m=nq)$

$$\hat{X}(s,\chi)=\hat{X}(s)\,\mathrm{e}^{-im\chi}$$
$$\hat{V}(s,\chi)=\hat{V}(s)\,\mathrm{e}^{-im\chi} \tag{5.83}$$

and the component $\hat{Z}$ is related to $\hat{V}$ by $\nabla\cdot\xi=0$. Since

$$r(s,\chi)=R+\varrho(s,\chi)\cos\chi \tag{5.84}$$

the angular variation of $\hat{Z}(s,\chi)$ shows dominant $m+1$ and $m-1$ poloidal mode numbers.

Experience shows that unstable modes are well represented numerically when this marginal solution is obtained with high accuracy. To obtain high accuracy, one possibility is to Fourier-expand the solution in the poloidal direction χ. On each integer nq surface, the mode can then exactly fulfil $F[\hat{V}]=F[\hat{X}]=0$ when it has to. In the radial direction one can choose finite elements belonging to a class satisfying $D=0$, for instance linear elements in s for $\hat{X}$ and piecewise constant elements in s for $\hat{V}$. This is the approach made in the PEST code (*Grimm* et al. 1976).

Another possibility is to choose the so-called finite hybrid elements as described in Chap. 4. We do not repeat the formulation here since, as a comparison between the expressions (5.52–56) and (4.4) shows, the expressions for the potential and kinetic energies W_p and W_k have the same structure. It is evident that the same finite hybrid elements may be used in the vacuum region, since W_v (5.79) and W_p (5.52) have the same structure.

5.6 Extraction of the Rapid Angular Variation

Since 1978 there has been a lively interest in the spectrum of internal modes with very high toroidal wave number. The so-called ballooning mode theory (*Connor* et al. 1978) provides a local stability criterion. This criterion depends on the local behavior of the pressure and current profiles, and on the geometrical curvature. Local changes in these quantities can strongly affect the stability of such modes. On the other hand, experimentalists believe that low-n ideal global MHD modes can be stabilized by the first wall which acts as a perfect conductor at least over the Alfvén time scale of the order of a microsecond.

It is generally accepted now that either low-n kink modes (especially $n=1$) or high-n ballooning modes give the stability limit. However, one knows that extremely high-n internal modes can be strongly affected by minor changes in the profiles and the geometry, or by the inclusion of additional physics, and can eventually be stabilized by such effects. The remaining modes are then those with intermediate n-numbers.

Using the approach discussed in the previous sections of the present chapter, the highest n-number treatable is given by $nq_p \approx 10$. This limit is due to the difficulty in reproducing numerically the fast angular variation of the mode (since it is very expensive with respect to computing time). From the ballooning mode theory we know that the major angular behavior of the displacement components $\hat{X}$, $\hat{V}$, and $\hat{Z}$ can be extracted by a variable transformation of the form

$$\begin{pmatrix} \bar{X}(s,\chi) \\ \bar{V}(s,\chi) \\ \bar{Z}(s,\chi) \end{pmatrix} = \begin{pmatrix} \hat{X}(s,\chi) \\ \hat{V}(s,\chi) \\ \hat{Z}(s,\chi) \end{pmatrix} \mathrm{e}^{-inq\chi} \,. \tag{5.85}$$

On $s=\text{const}$ surfaces with large nq values, $\hat{X}$, $\hat{V}$, and $\hat{Z}$ are strongly oscillating, whereas $\bar{X}$, $\bar{V}$, and $\bar{Z}$ are slowly varying functions of χ. In general, a fast radial variation remains, due to the many singular surfaces within the plasma. If we replace $\hat{X}$, $\hat{V}$, and $\hat{Z}$ by $\bar{X}$, $\bar{V}$, and $\bar{Z}$ in (5.52–56, 79) only the integrands I_1 to I_5 are changed. They become

$$\begin{aligned} I_1 &= \frac{\partial \bar{X}}{\partial \chi} \\ I_2 &= \bar{D} = \frac{\partial \bar{X}}{\partial s} + \frac{\partial \bar{V}}{\partial \chi} - inq\bar{V} - in\frac{dq}{ds}\chi\bar{X} \\ I_3 &= \bar{D} + H\bar{X} + \beta_\chi \frac{\partial \bar{X}}{\partial \chi} - \frac{\partial \bar{V}}{\partial \chi} \\ I_4 &= \bar{D} + \bar{X}\frac{\partial \ln r^2}{\partial s} + \bar{V}\frac{\partial \ln r^2}{\partial \chi} + \frac{1}{r^2}\frac{\partial \bar{Z} r^2}{\partial \chi} \\ I_5 &= \hat{X} \,. \end{aligned} \tag{5.86}$$

Note that for the case of large nq values all dominant terms are in D which only appears in positive quadratic terms. For unstable modes and large nq values the dominant terms in D have to cancel each other. This cancellation cannot be achieved with enough accuracy by discretizing X and V with regular finite elements. As a consequence, no unstable modes with a reasonable number of mesh cells can be found. However, with a finite hybrid element representation, the cancellation can occur exactly and the unstable modes can be found.

The eigenfunction $\xi(s,\chi)$ is 2π periodic, and if up-down symmetry is imposed;

$$\begin{aligned}\hat{X}(s,\chi) &= -\hat{X}^*(s,-\chi)\\ \hat{V}(s,\chi) &= \hat{V}^*(s,-\chi)\\ \hat{Z}(s,\chi) &= \hat{Z}^*(s,-\chi)\ ,\end{aligned} \tag{5.87}$$

where * denotes the complex conjugate. The symmetry conditions (5.87) can be separated into the boundary conditions

$$\begin{aligned}\hat{X}^{\mathrm{R}}(s,0) &= \hat{X}^{\mathrm{R}}(s,\pi)=0\\ \hat{V}^{\mathrm{I}}(s,0) &= \hat{V}^{\mathrm{I}}(s,\pi)=0\\ \hat{Z}^{\mathrm{I}}(s,0) &= \hat{Z}^{\mathrm{I}}(s,\pi)=0\end{aligned} \tag{5.88}$$

and into natural conditions at $\chi=0$ and $\chi=\pi$ on $\hat{X}^{\mathrm{I}}$, $\hat{V}^{\mathrm{R}}$, and $\hat{Z}^{\mathrm{R}}$ which do not have to be imposed. The transformed components $\bar{X}$, $\bar{V}$, and $\bar{Z}$ fulfil the same symmetry conditions at $\chi=0$ as $\hat{X}$, $\hat{V}$, and $\hat{Z}$. At $\chi=\pi$, however, the symmetry conditions change drastically since $\bar{X}$, $\bar{V}$, and $\bar{Z}$ are not 2π periodic. Introducing two angular positions at $\chi=\pi-\varepsilon$ and at $\chi=\pi+\varepsilon$, and recalling the relation (5.85), the symmetry conditions (5.87) around $\chi=\pi$ become;

$$\begin{aligned}
&\bar{X}^{\mathrm{R}}(\psi,\pi+\varepsilon)\cos nq(\pi+\varepsilon)-\bar{X}^{\mathrm{I}}(\psi,\pi+\varepsilon)\sin nq(\pi+\varepsilon)\\
&\quad=-\bar{X}^{\mathrm{R}}(\psi,\pi-\varepsilon)\cos nq(\pi-\varepsilon)+\bar{X}^{\mathrm{I}}(\psi,\pi-\varepsilon)\sin nq(\pi-\varepsilon)\\
&\bar{X}^{\mathrm{R}}(\psi,\pi+\varepsilon)\sin nq(\pi+\varepsilon)+\bar{X}^{\mathrm{I}}(\psi,\pi+\varepsilon)\cos nq(\pi+\varepsilon)\\
&\quad=\bar{X}^{\mathrm{R}}(\psi,\pi-\varepsilon)\sin nq(\pi-\varepsilon)+\bar{X}^{\mathrm{I}}(\psi,\pi-\varepsilon)\cos nq(\pi-\varepsilon)\\
&\bar{V}^{\mathrm{R}}(\psi,\pi+\varepsilon)\cos nq(\pi+\varepsilon)-\bar{V}^{\mathrm{I}}(\psi,\pi+\varepsilon)\sin nq(\pi+\varepsilon)\\
&\quad=\bar{V}^{\mathrm{R}}(\psi,\pi-\varepsilon)\cos nq(\pi-\varepsilon)-\bar{V}^{\mathrm{I}}(\psi,\pi-\varepsilon)\sin nq(\pi-\varepsilon)\\
&\bar{V}^{\mathrm{R}}(\psi,\pi+\varepsilon)\sin nq(\pi+\varepsilon)+\bar{V}^{\mathrm{I}}(\psi,\pi+\varepsilon)\cos nq(\pi+\varepsilon)\\
&\quad=-\bar{V}^{\mathrm{R}}(\psi,\pi-\varepsilon)\sin nq(\pi-\varepsilon)-\bar{V}^{\mathrm{I}}(\psi,\pi-\varepsilon)\cos nq(\pi-\varepsilon)\\
&\bar{Z}^{\mathrm{R}}(\psi,\pi+\varepsilon)\cos nq(\pi+\varepsilon)-\bar{Z}^{\mathrm{I}}(\psi,\pi+\varepsilon)\sin nq(\pi+\varepsilon)\\
&\quad=\bar{Z}^{\mathrm{R}}(\psi,\pi-\varepsilon)\cos nq(\pi-\varepsilon)-\bar{Z}^{\mathrm{I}}(\psi,\pi-\varepsilon)\sin nq(\pi-\varepsilon)\\
&\bar{Z}^{\mathrm{R}}(\psi,\pi+\varepsilon)\sin nq(\pi+\varepsilon)+\bar{Z}^{\mathrm{I}}(\psi,\pi+\varepsilon)\cos nq(\pi+\varepsilon)\\
&\quad=-\bar{Z}^{\mathrm{R}}(\psi,\pi-\varepsilon)\sin nq(\pi-\varepsilon)-\bar{Z}^{\mathrm{I}}(\psi,\pi-\varepsilon)\cos nq(\pi-\varepsilon)\ .
\end{aligned} \tag{5.89}$$

Introducing the variable transformation

$$\begin{pmatrix} \bar{X}^{R}(\psi,\pi-\varepsilon) \\ \bar{X}^{I}(\psi,\pi-\varepsilon) \\ \bar{X}^{R}(\psi,\pi+\varepsilon) \\ \bar{X}^{I}(\psi,\pi+\varepsilon) \end{pmatrix} = \begin{pmatrix} 1 & 0 & 0 & 0 \\ 0 & 1 & 0 & 0 \\ -\cos 2nq\pi & -\sin 2nq\pi & 1 & 0 \\ \sin 2nq\pi & -\cos 2nq\pi & 0 & 1 \end{pmatrix} \begin{pmatrix} \bar{X}^{R}(\psi,\pi-\varepsilon) \\ \bar{X}^{I}(\psi,\pi-\varepsilon) \\ \bar{X}^{R}(\psi,\pi+\varepsilon) \\ \bar{X}^{I}(\psi,\pi+\varepsilon) \end{pmatrix} \tag{5.90}$$

and

$$\begin{pmatrix} \bar{Y}^{R}(\psi,\pi-\varepsilon) \\ \bar{Y}^{I}(\psi,\pi-\varepsilon) \\ \bar{Y}^{R}(\psi,\pi+\varepsilon) \\ \bar{Y}^{I}(\psi,\pi+\varepsilon) \end{pmatrix} = \begin{pmatrix} 1 & 0 & 0 & 0 \\ 0 & 1 & 0 & 0 \\ \cos 2nq\pi & -\sin 2nq\pi & 1 & 0 \\ -\sin 2nq\pi & -\cos 2nq\pi & 0 & 1 \end{pmatrix} \begin{pmatrix} \bar{Y}^{R}(\psi,\pi-\varepsilon) \\ \bar{Y}^{I}(\psi,\pi-\varepsilon) \\ \bar{Y}^{R}(\psi,\pi+\varepsilon) \\ \bar{Y}^{I}(\psi,\pi+\varepsilon) \end{pmatrix}, \tag{5.91}$$

where Y is either V or Z, the expressions in (5.89) are identical to

$$\begin{aligned} \bar{X}^{R}(\psi,\pi+\varepsilon) &= \bar{X}^{I}(\psi,\pi+\varepsilon) = \bar{V}^{R}(\psi,\pi+\varepsilon) = \bar{V}^{I}(\psi,\pi+\varepsilon) \\ &= \bar{Z}^{R}(\psi,\pi+\varepsilon) = \bar{Z}^{I}(\psi,\pi+\varepsilon) = 0 \ . \end{aligned} \tag{5.92}$$

The initial problem (5.35, 52, 55, and 79) discretized by finite hybrid elements leads to an eigenvalue problem

$$A\hat{\boldsymbol{x}} = \omega^2 B\hat{\boldsymbol{x}} \ , \quad \text{where} \tag{5.93}$$

$$\hat{\boldsymbol{x}} = (\hat{X}_0, \hat{V}_0, \hat{Z}_0, \dots, \hat{X}_j, \hat{V}_j, \hat{Z}_j, \dots, \hat{X}_N, \hat{V}_N, \hat{Z}_N) \ . \tag{5.94}$$

The variable transformation (5.85) together with (5.90) and (5.91) can be written as

$$\hat{\boldsymbol{x}} = U\bar{\boldsymbol{x}} \tag{5.95}$$

where U is the transformation matrix. Multiplying (5.93) with U^{T} from the left-hand side gives a new eigenvalue problem for the transformed vector $\bar{\boldsymbol{x}}$

$$\bar{A}\bar{\boldsymbol{x}} = \omega^2 \bar{B}\bar{\boldsymbol{x}} \ , \quad \text{where} \tag{5.96}$$

$$\bar{A} = U^{T}AU \quad \text{and} \tag{5.97}$$

$$\bar{B} = U^{T}BU \tag{5.98}$$

are symmetric again, and B is positive definite.

5.7 Calculation of β-Limits (with F. Troyon)

One of the applications of the ERATO code is the computation of the limiting plasma pressure above which the plasma is unstable. This limiting pressure scales with the magnetic pressure so that it is best expressed as a limit on β, defined as

$$\beta = \frac{\int\limits_{\Omega_p} p\,dr\,dz}{\frac{1}{2\mu_0}\int\limits_{\Omega_p} B^2 dr\,dz}\,, \tag{5.99}$$

which is the ratio between the average plasma pressure and the average magnetic pressure. It is believed that β-values of the order of 6% are necessary for a commercially viable tokamak reactor. This maximum value of β for which there is stability depends on a number of parameters, plasma shape, current and pressure profiles.

As an example of such an application, we show here a typical case of limited parametric optimisation of a configuration. The shape of the plasma cross section is given by

$$\begin{aligned} r &= R_0 + a\cos(\theta' + d\sin\theta') \\ z &= Ea\sin\theta'\ . \end{aligned} \tag{5.100}$$

The parameters are the major plasma radius R_0, the minor plasma radius a, the elongation E, and the triangularity d. The JET plasma shape is well represented by (5.100) if one chooses $R_0 = 2.96m$, $a = 1.25m$, $d = 0.3$, and $E = 1.68$, while the symmetric INTOR reference shape (Phase 1) corresponds to $R_0 = 5.2m$, $a = 1.3m$, $d = 0.3$, and $E = 1.6$.

In the source terms the simple polynomial expressions

$$\begin{aligned} p(\psi) &= p_1\psi^2 + p_2\psi^3 \\ T^2(\psi) &= T_{\mathrm{lim}}^2 + t_1\psi^2\ , \end{aligned} \tag{5.101}$$

where T_{lim} is the toroidal flux at the limiter and t_1, p_1, and p_2 are three free parameters, give sufficient freedom to adjust the safety factor on axis q_0, the poloidal beta β_p and the total plasma current I.

In our study we fix the total plasma current

$$I = \int\limits_{\Omega_p} \left(\frac{dp}{d\psi} r + \frac{1}{2r}\frac{dT^2}{d\psi}\right) dr\,dz \tag{5.102}$$

which determines one of the three parameters p_1, p_2 or t_1 in (5.101). In the β-optimization procedure, we can fix the parameter p_1, – this determines λ in

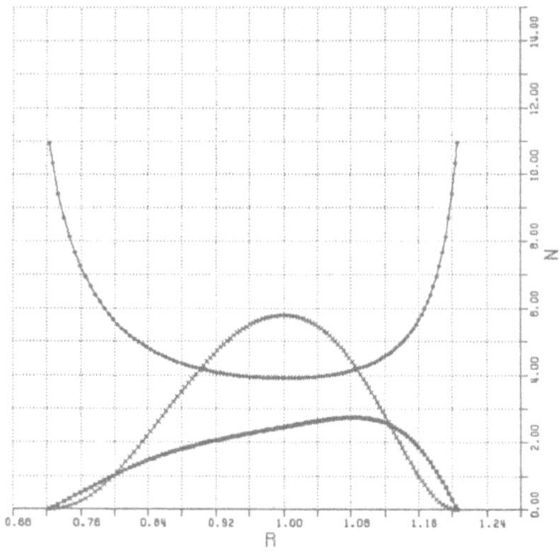

Fig. 5.6. Pressure, current and q-profiles in the symmetry plane for INTOR. $I = 5.9$ MA, $p = 20$, $t = 0.10$, $q_0 = 1.501$, $q_p = 3.567$

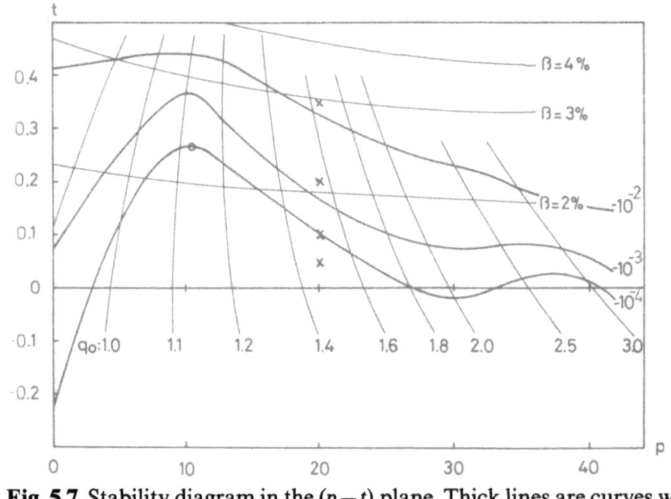

Fig. 5.7. Stability diagram in the $(p-t)$ plane. Thick lines are curves with constant growthrate squared $\omega^2 = -10^{-4}$, -10^{-3}, and -10^{-2} for the $n = 1$ external kink mode. Curves are also plotted of constant $\beta = 2\%$, 3%, and 4% and of constant $q_0 = 0.9$, 1.0, 1.1, 1.2, 1.4, 1.6, 1.8, 2.0, 2.5, and 3.0. The curve for $\Gamma^2 = 10^{-4}$ is considered to be the stability limit. The maximum stable β point is marked with 0. ($\times$ $\times$ $\times$) are snap shots discussed in Figs. 5.8,9

(5.22)– and search for those values of p_2 and t_1 which give the highest value of β for which the plasma is still stable.

For an optimized configuration, Fig. 5.6 shows the profiles of the pressure $p(r)$, safety factor $q(r)$ and current density $j(r)$ across the equatorial plane ($z = 0$), for an INTOR-shaped plasma with $I = 5.9$ MA and a toroidal magnetic field at the magnetical axis of 5.5 Tesla. The parameters for this case are $p = p_2/p_1 = 20$ and $t = t_1/p_1 = 0.10$. The corresponding β is 2%.

The details of the optimisation are shown in Fig. 5.7. Each point in the plane (p, t) represents an equilibrium. The corresponding values of q_0 and β can also be read in the same plane since to each point correspond single values of q_0 and β. Plotted in this figure are the lines of constant normalized growth rate squared, Γ^2, of the $n=1$ free boundary unstable mode (no shell around the plasma). The curves are computed at a fixed resolution of 60 radial and 60 azimuthal intervals over the full cross section. The ideal MHD model cannot be expected to hold at arbitrary slow growthrates as other phenomena become important. It is estimated that a residual growthrate of $\Gamma^2=10^{-4}$, which corresponds to growthtimes of the order of 100 µs for typical tokamak parameters, must be close to the upper limit of credibility of the model.

We choose this limit of $\Gamma^2=10^{-4}$ as the stability limit. This procedure has been proposed by *Goedbloed* and *Sakanaka* (1974) on the same physical grounds. There is also a numerical argument which has led to the choice of $\Gamma^2=10^{-4}$ as the stability limit. It is based on considerations of the speed of convergence. To obtain the true Γ^2 it is necessary to extrapolate the computed Γ^2 to zero mesh size. Figure 5.8 shows four examples of such a convergence study. The parameters of these equilibria, labeled with crosses in Fig. 5.7, are

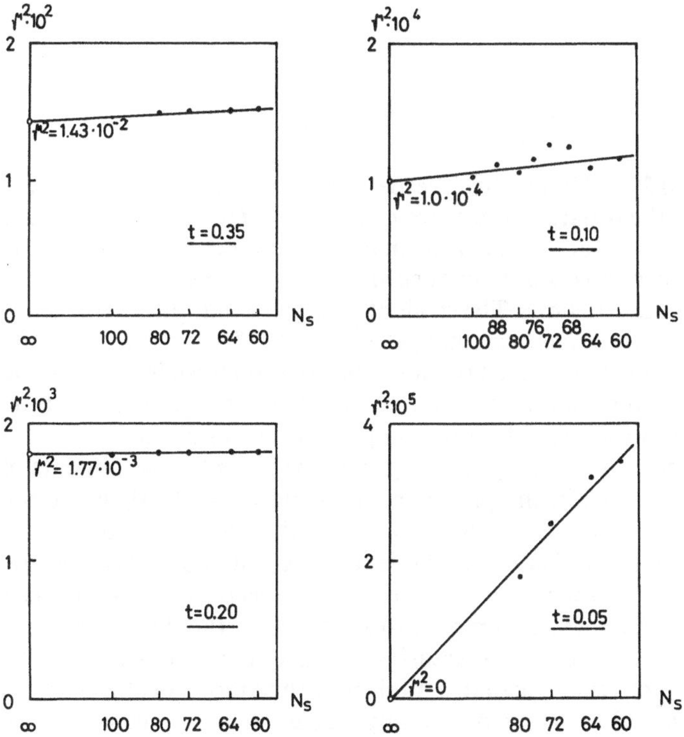

Fig. 5.8. Convergence curves for $p=20$ and $t=0.35$, 0.20, 0.10, and 0.05 (crosses in Fig. 5.7). Cases for $t=0.35$ and $t=0.20$ are unstable, marginal for $t=0.10$ and stable for $t=0.05$

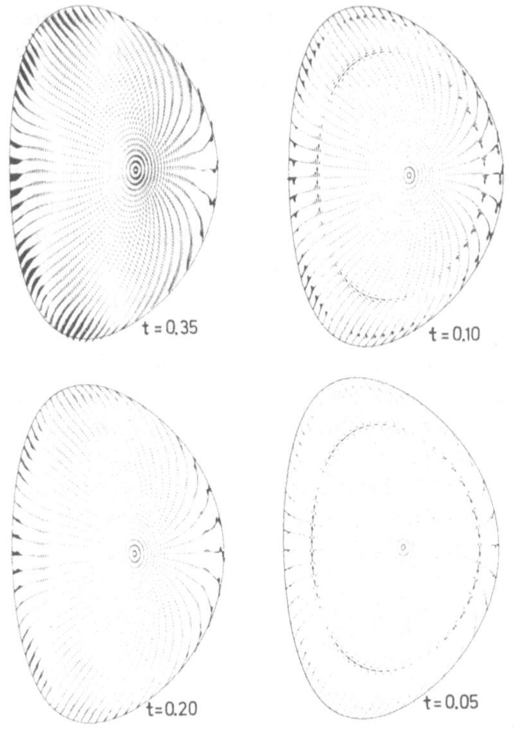

Fig. 5.9. Displacement patterns for the modes studied in Fig. 5.7 using a 60×60 non-equidistant grid

$p=20$ and $t=0.35$, 0.20, 0.10, and 0.05, respectively. The growthrate squared, Γ^2, is plotted versus the square of the average mesh size. The ratio between the number of intervals in the radial and azimuthal directions is kept constant. The radial mesh is dense around each singular surface to resolve the rapid change of the eigenfunction in these regions. The straight lines show that the convergence is quadratic. The experience has been that, when $\Gamma^2 < 10^{-4}$ at a resolution of 60×60 intervals in the whole domain, it is impossible to distinguish from the value $\Gamma^2 = 0$, which is the lower limit of the continuum. The limited resolution of the numerical equilibrium used also becomes important in this range.

Typical eigenvectors are displayed in Fig. 5.9. The poloidal components of the eigenvectors are shown in the particular meridian plane $\phi = 0$, where the mode is up-down symmetric. Again we fix $p=20$ and vary $t=0.35$, 0.20, 0.10, and 0.05. The most striking feature is the increasing radial structure on the singular surfaces, which appears as the growthrate decreases. For the last equilibrium, with $t=0.05$, which is most probably stable, the displacement is localized around the $q=2$ surface. Increasing the resolution from the 60×60 mesh used for these plots would lead to a mode more and more localized on the $q=2$ surface, corresponding to the marginally stable "singular eigenmode" of the lower edge of the continuous spectrum. Similar studies are reported in *Gruber* et al. (1981c).

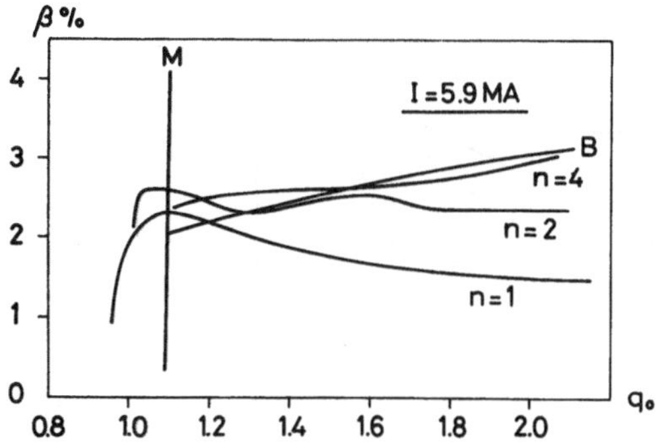

Fig. 5.10. Stability limits in the $(q_0-\beta)$ plane. The curve labelled with $n=1$ is a cross plot of Fig. 5.7. The curves labelled with M and B are the Mercier and the ballooning limits. Stability is below and on the right. From *Troyon* et al. (1983)

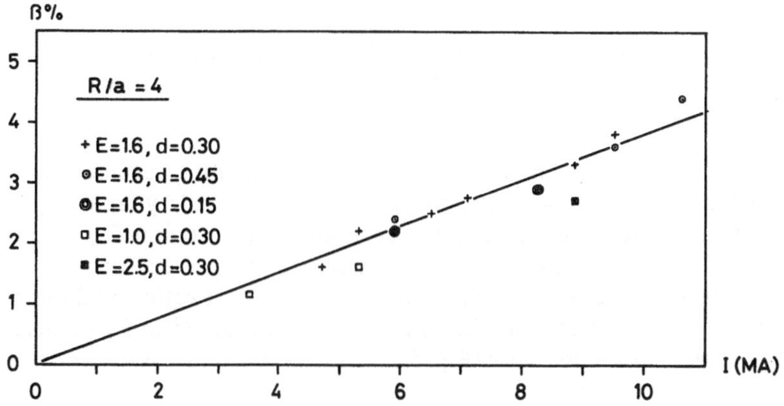

Fig. 5.11. Dependence of the maximum β on the plasma current I for INTOR. From *Troyon* et al. (1983)

The curve for $\Gamma^2=10^{-4}$ in Fig. 5.7 is cross-plotted and shown again in Fig. 5.10 in the plane (β, q_0). In this figure are also represented the Mercier limit (M), the ballooning limit (B) and the 10^{-4} stability limits of the free boundary modes $n=2$ and $n=4$. It is clearly seen that the stability limit is either given by the external mode $n=1$ or by the ballooning limit corresponding to $n=\infty$. The ballooning limit is sensitive to local changes in the pressure gradient. By slightly redistributing the pressure it is possible to shift the ballooning limit above the $n=1$ free boundary limit, as long as the Mercier criterion is satisfied. Ideally, it must be possible to have the ballooning limit smoothly joining the Mercier limit above the crossing point between the Mercier limit and the $n=1$ free boundary limit. It means that the highest value of stable β is either the intersection of the Mercier limit, with the $n=1$ limit, or the maximum point of the $n=1$ external kink curve.

This optimization procedure can be repeated for different values of the total current at different elongations and triangularities. The maximum values of β at each current are shown in Fig. 5.11. One sees that the maximum value of β increases roughly linearly with the total current. There is a small oscillation

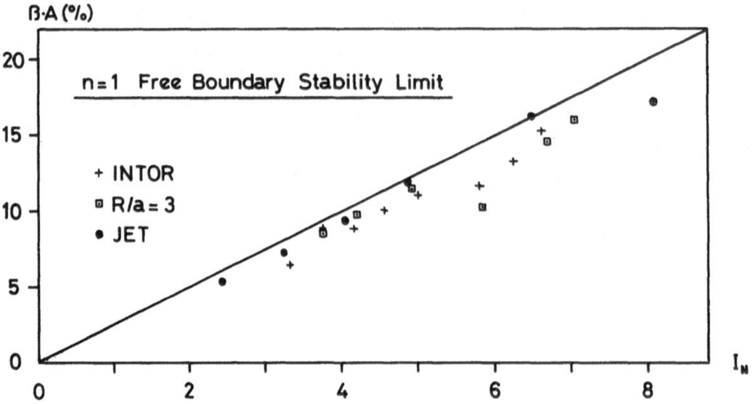

Fig. 5.12. Dependence of the limiting $\beta \cdot A$ given by the $n=1$ free boundary mode on the normalized current I_n. From *Troyon* et al. (1983)

around the straight line with a maximum each time a constant q surface enters the plasma surface. At the highest current of $I=12$ MA, the safety factor q_s at the plasma surface is just above 2. At higher currents q_s drops below 2, and the configuration is found to be always unstable with the profiles chosen.

These calculations have been repeated for different aspect ratios, including the JET set of parameters. All the maximum values of β obtained so far in these studies are collected in a single graph (Fig. 5.12) using as variables $\beta * A$ and the normalized current

$$I_N = \frac{\mu_0 I A^2}{T_{\text{lim}}} , \tag{5.103}$$

where I is the total current in Amperes, $A = R_0/a$ and T_{lim} is the toroidal flux in teslameters at the limiter. All points fall below an upper limit shown as straight line in Fig. 5.12. This limit can be written as (Troyon limit)

$$\max(\beta) = 2.5 \mu_0 \frac{IA}{T_{\text{lim}}} , \tag{5.104}$$

restricted by the condition that

$$q_p \gtrsim 2 . \tag{5.105}$$

This means that for a given total current, β cannot be increased by altering elongation or triangularity at constant current. However, it is possible to increase the critical total current (given by $q_p \sim 2$) by increasing the elongation or the triangularity.

The simplicity of this result invites a comparison with the available experimental information. This is done in Fig. 5.13 where the highest values of

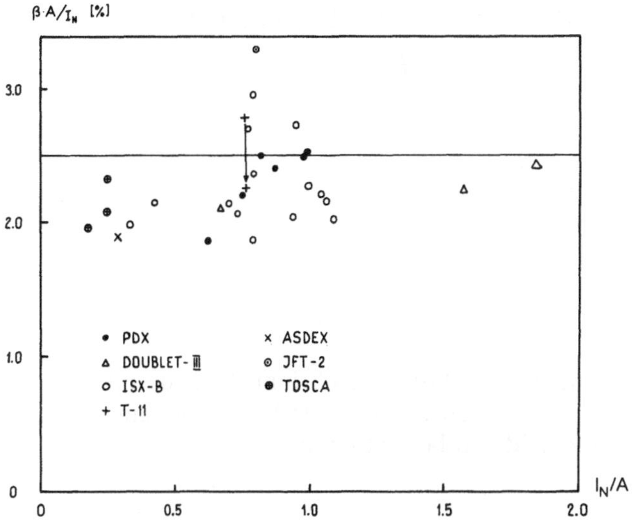

Fig. 5.13. Comparison of the numerically obtained limiting β with experimental optimal β values measured on different tokamaks. From *Troyon* and *Gruber* (1984)

$\beta A/I_N$ achieved in various conditions in PDX (*Johnson* et al. 1983), Doublet III (*Burrell* et al. 1983; *Nagami* et al. 1983), ISX-B (*Swain* et al. 1981; *Neilson* et al. 1983; *Charlton* et al. 1984), T–11 (*Leonov* et al. 1981; *Barsukov* et al. 1983), Asdex (*Wagner* et al. 1983), JFT-2 (*Yamamoto* et al. 1981; *Suzuki* et al. 1981) and Tosca (*Morris* and *Todd* 1983) are shown together with the upper limit of β (5.104). There are five values of β coming from ISX-B, (*Swain* et al. 1981), JFT-2 (*Suzuki* et al. 1981), and T-11 (*Leonov* et al. 1981) which overcome our limit domain of β. All these values were measured before 1981. All the more recent limit values of β found with more diagnostics (for example, diamagnetic loop measurements) lie within the numerically determined region. It must be noted here that only β-limit points for which $\beta A/I_N > 1.9$ are represented in this diagram.

We can claim that the experimental limit values of β have not convincingly overcome our numerically determined limit domain of β, in spite of the obvious limitations of the optimisation carried out with only a few parameters. More details on these optimisation calculations can be found in *Troyon* et al. (1983) and in *Troyon* and *Gruber* (1984).

6. HERA: Application to Helical Geometry (Peter Merkel, IPP Garching)

6.1 Equilibrium

A helically symmetric magnetohydrostatic equilibrium depends on the two variables r and $\zeta = \phi - hz$, where (r, ϕ, z) are cylindrical coordinates. The helically symmetric magnetic field can be written as

$$\begin{aligned} \boldsymbol{B} &= T\boldsymbol{u} + \boldsymbol{u} \times \nabla\psi \\ \boldsymbol{u} &= \frac{1}{1+r^2h^2}(hr\boldsymbol{e}_\phi + \boldsymbol{e}_z) \ , \end{aligned} \tag{6.1}$$

where T and ψ are arbitrary functions of r and ζ, and $\boldsymbol{e}_\phi$, $\boldsymbol{e}_z$ are the unit vectors in the ϕ and z direction, respectively. The helical pitch is defined by $2\pi/h$.

The equilibrium equation is given by $(' \equiv d/d\psi)$

$$\begin{aligned} &\frac{1}{r}\frac{\partial}{\partial r}\left(r|\boldsymbol{u}|^2\frac{\partial\psi}{\partial r}\right) + \frac{1}{r^2}\frac{\partial^2\psi}{\partial\zeta^2} = -j|\boldsymbol{u}| - 2h|\boldsymbol{u}|^4 T \\ &|\boldsymbol{u}|j = p' + TT'|\boldsymbol{u}|^2 \ , \qquad |\boldsymbol{u}|^2 = \frac{1}{1+r^2h^2} \ . \end{aligned} \tag{6.2}$$

As for the axisymmetric case, the pressure p and the longitudinal field T are functions of the flux function ψ: $p = p(\psi)$, $T = T(\psi)$; and j is the current density along the helix.

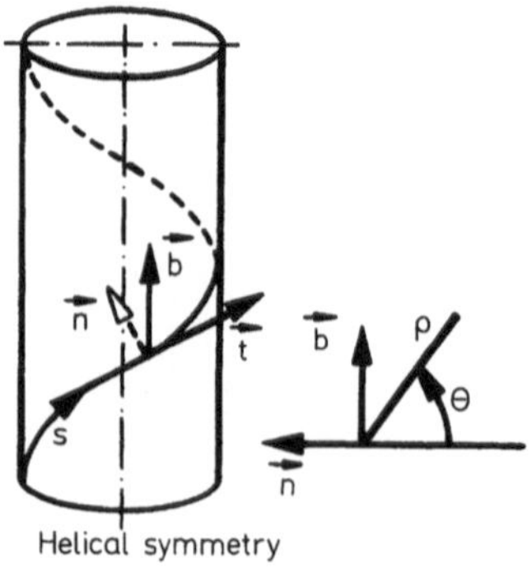

Fig. 6.1. The (s, θ, ϱ) coordinate system used in the HASE equilibrium code; s is the arclength along the magnetic axis, ϱ the distance from the magnetic axis, θ the poloidal angle in a plane perpendicular to the magnetic axis

The HERA stability code is applied mainly to numerically calculated equilibria. These equilibria are calculated with the HASE code (*Merkel* and *Nührenberg* 1984) which solves (6.2). The partial differential equation which is non-linear in general, is not treated as a boundary value problem, but by the method of expansion in aspect ratio to arbitrary order such that (see Fig. 6.1)

$$\psi(\varrho,\theta)=\sum_{m=2}^{\infty}\sum_{n=0}^{m} S_{mn}\varrho^m \cos n\theta \ . \tag{6.3}$$

Here, ϱ, θ are polar coordinates in a plane normal to the magnetic axis, which is defined by its curvature

$$\kappa=\frac{R_0h^2}{1+R_0^2h^2} \tag{6.4}$$

and its torsion

$$\tau=\frac{h}{1+R_0^2h^2} \ , \tag{6.5}$$

where R_0 is the distance of the magnetic axis from the z-axis. The arbitrary profile functions $p'(\psi)$ and $T(\psi)$ are chosen as polynomials in ψ

$$\begin{aligned} T(\psi)&=\sum_{n=0}^{n_f} f_n\psi^n \\ p'(\psi)&=\sum_{n=0}^{n_p} p_n\psi^n \ . \end{aligned} \tag{6.6}$$

The equilibrium is completely determined by prescribing the magnetic axis, the values f_n, p_n and the diagonal coefficients S_{nn}.

Introducing the half-axis ratio E of the cross section and the rotational transform ι_0 at the magnetic axis, the coefficients S_{20}, S_{22} and f_1 become

$$\begin{aligned} S_{20}&=\frac{1}{2}\left(E+\frac{1}{E}\right) \\ S_{22}&=\frac{1}{2}\left(E-\frac{1}{E}\right) \\ f_1&=-p_0-\frac{1+\iota_0}{2}\left[\left(E+\frac{1}{E}\right)(1+\iota_0)-2\tau\right] \ . \end{aligned} \tag{6.7}$$

The length and the magnetic field are normalized by setting

$$\begin{aligned} &\kappa^2+\tau^2=1 \\ &T(\varrho=0,\theta)=1 \ . \end{aligned} \tag{6.8}$$

This normalization fixes

$$R_0^2=(h^2-1)/h^2 \tag{6.9}$$

and implies that $h^2 \geqq 1$. For $h=1$, the magnetic axis falls on the helical axis which leads to $R_0=0$, $\kappa=0$ and $\tau=1$. For $h=\infty$, $\tau=0$ and $\kappa=1$, one obtains the axisymmetric limit.

Then, to determine an equilibrium, the following quantities have to be prescribed:

κ	curvature of magnetic axis
E	half-axes ratio of cross-section near the magnetic axis.
ι_0	rotational transform at the magnetic axis
$p_n, n \geqq 0$	coefficients of pressure function
$f_n, n \geqq 2$	coefficients of longitudinal field functions
$S_{nn}, n \geqq 3$	diagonal coefficients of power series .

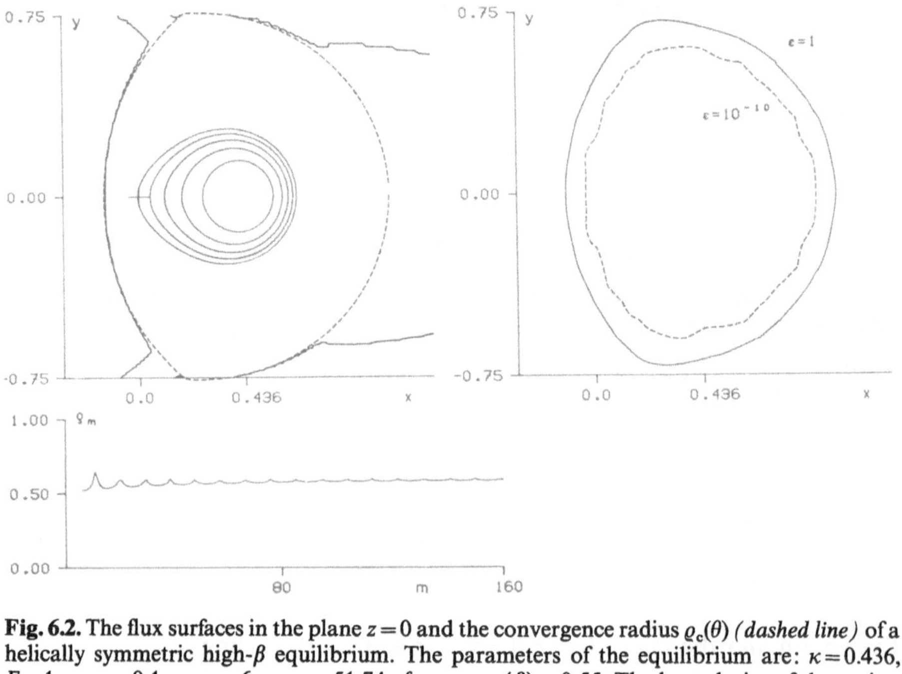

Fig. 6.2. The flux surfaces in the plane $z=0$ and the convergence radius $\varrho_c(\theta)$ *(dashed line)* of a helically symmetric high-β equilibrium. The parameters of the equilibrium are: $\kappa=0.436$, $E=1$, $\iota_0=-0.1$, $p_0=-6$, $p_1=-51.74$, $f_1=-p_0$, $\langle\beta\rangle=0.55$. The boundaries of the region are shown, in which the error ε of the solution becomes $\varepsilon<10^{-10}$ (– – – –) and $\varepsilon<1$ (– – – –). The last plot shows the convergence of the solution for $\theta=0$; $\varrho_m(0)$ is plotted versus the ordering number m

The coefficients S_{nn} determine the form of the flux surfaces; (S_{22}-ellipticity, S_{33}-triangularity, S_{44}-quadrangularity, ...). If the series (6.3) converges, then $\psi(\varrho, \theta)$ is an equilibrium solution. The convergence radius $\varrho_c(\theta)$ is defined as

$$\varrho_c(\theta) = \lim_{m\to\infty} \varrho_m(\theta) \quad \text{with} \quad \varrho_m(\theta) = \left| \sum_{n=0}^{m} S_{mn} \cos n\theta \right|^{-1/m} . \tag{6.10}$$

An analytical proof of convergence is not possible in general. Therefore, having calculated the coefficients S_{mn} and the $\varrho_m(\theta)$ up to values $m=M$, where M reached values of $M=160$ in some cases, the convergence radius was determined numerically. In all cases considered, the $\varrho_m(\theta)$ reached its asymptotic value for $m<M$ with good approximation (Fig. 6.2).

6.2 Variational Formulation of the Stability Problem

As for the axisymmetric case we start from the variational form, (5.35–38) subject to (5.39–41). For a helically symmetric geometry the independent variables (Fig. 6.3) are (ψ, χ, z), where the "poloidal" variable χ is determined by requiring that the Jacobian J is

$$J \equiv \frac{1}{\nabla z \cdot (\nabla\psi \times \nabla\chi)} = \frac{1}{|\boldsymbol{u}|^2 g(\psi)} , \tag{6.11}$$

where $g(\psi)$ is a function of ψ only. From this follows the differential equation for χ:

$$d\chi = \boldsymbol{dl} \cdot \nabla\chi = \boldsymbol{dl} \cdot (\boldsymbol{u} \times \nabla\psi)/(J|\boldsymbol{u}|^2 |\nabla\psi|^2) , \tag{6.12}$$

where $\boldsymbol{dl}$ is the line element. The value of $J|\boldsymbol{u}|^2$ on each flux surface is obtained by normalizing χ to $0<\chi<2\pi$ for one period.

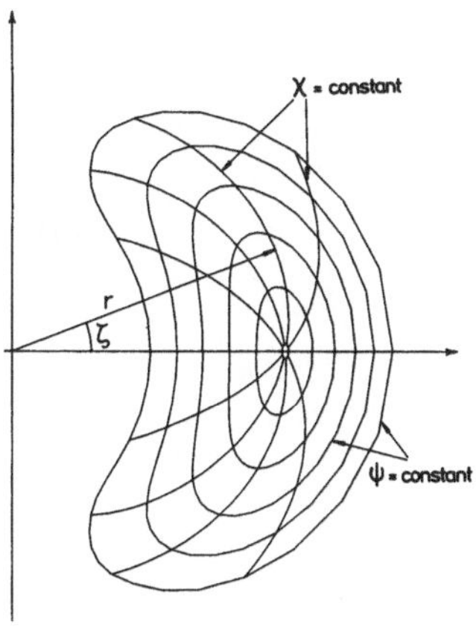

Fig. 6.3. Coordinate system in a z=constant plane for a helically symmetric configuration (r, ζ) and the flux coordinate system (ψ, χ)

As in the axisymmetric case the displacement vector ξ is Fourier-analysed with respect to time and the ignorable coordinate z

$$\xi = \xi_k(r, \zeta)\, e^{i\omega t + ikz} , \tag{6.13}$$

and the solutions with different values of k decouple. In the plasma region ξ_k is represented by the components X_k, V_k, Z_k as follows

$$\xi_k = \frac{X_k}{|\boldsymbol{u}|^2}(\nabla\chi \times \boldsymbol{u}) + \frac{V_k}{|\boldsymbol{u}|^2}(\boldsymbol{u} \times \nabla\psi) + \frac{Z_k}{|\boldsymbol{u}|^2}\boldsymbol{B} . \tag{6.14}$$

Introducing the same variable transformations as in (5.50), i.e.,

$$\begin{aligned} \hat{X} &= X_k \\ \hat{V} &= 2s\psi_p V_k \\ \hat{Z} &= 2s\psi_p Z_k \end{aligned} \tag{6.15}$$

and the independent variable $s = \sqrt{(\psi_p + \psi)/\psi_p}$, the potential energy W_p can be written as (*Gruber* et al. 1981b)

$$W_p = \frac{1}{2}\int_0^1\int_0^{2\pi} ds\, d\chi \{a_1|I_1|^2 + a_2|I_2|^2 + a_3|I_3|^2 + a_4|I_4|^2 - a_5|I_5|^2\} ,$$

where

$$\begin{aligned} I_1 &= F[\hat{X}] = \frac{\partial \hat{X}}{\partial \chi} + ik\tilde{q}\hat{X} \\ I_2 &= D - 4s\psi_p h|\boldsymbol{u}|^2 \hat{X}/T \\ I_3 &= D + H\hat{X} - F[\hat{V}] + \beta_\chi F[\hat{X}] \\ I_4 &= D + \hat{X}\frac{\partial \ln 1/|\boldsymbol{u}|^2}{\partial s} + \hat{V}\frac{\partial \ln 1/|\boldsymbol{u}|^2}{\partial \chi} + |\boldsymbol{u}|^2 F\left[\frac{\hat{Z}}{|\boldsymbol{u}|^2}\right] \\ I_5 &= \hat{X} \\ D &= \frac{\partial \hat{X}}{\partial s} + \frac{\partial \hat{V}}{\partial \chi} + ik\alpha\hat{V} + (\beta_z - \alpha\beta_\chi)ik\hat{X} \end{aligned}$$

and (6.16)

$$\begin{aligned} H &= J|\boldsymbol{u}|^2 \frac{\partial(1/J|\boldsymbol{u}|^2)}{\partial s} + \frac{2s\psi_p j}{|\boldsymbol{u}|\,|\nabla\psi|^2} \\ a_1 &= \frac{2\psi}{J^3|\boldsymbol{u}|^4|\nabla\psi|^2 s} \\ a_2 &= \frac{T^2}{2J|\boldsymbol{u}|^2 s\psi_p} \end{aligned}$$

$$a_3 = \frac{|\nabla\psi|^2}{2J|\boldsymbol{u}|^2 s\psi_p}$$

$$a_4 = \frac{\gamma p}{2J|\boldsymbol{u}|^4 s\psi_p}$$

$$a_5 = \frac{4\psi}{J|\boldsymbol{u}|^4 s}\left[p'\left(\frac{\partial \ln|\boldsymbol{u}|}{\partial\psi}\right)_{\boldsymbol{n}} + \frac{j^2}{|\nabla\psi|^2} + \frac{j|\boldsymbol{u}|}{2|\nabla\psi|^2}\left(\frac{\partial|\nabla\psi|^2}{\partial\psi}\right)_{\boldsymbol{n}}\right.$$
$$\left. + h\left(|\boldsymbol{u}|^4 T' + j\frac{|\boldsymbol{u}|^3 T}{|\nabla\psi|^2}\right)\right]$$

with

$$j = p'/|\boldsymbol{u}| + TT'|\boldsymbol{u}|$$

$$\tilde{q} = J|\boldsymbol{u}|^2 T + \alpha$$

$$\alpha = J\nabla\psi\cdot(\nabla z \times \boldsymbol{u})$$

$$\beta_\chi = \nabla\psi\cdot\nabla\chi/|\nabla\psi|^2$$

$$\beta_z = \nabla z\cdot\nabla\psi/|\nabla\psi|^2$$

and ψ_p is the total flux in Ω_p. The rotational transform at the magnetic axis, ι_0, introduced in (6.7) is related to q_0 by

$$1 + \iota_0 = -\frac{1}{hq_0}\ . \tag{6.17}$$

The expression of the kinetic energy is

$$W_k = -\frac{\omega^2}{2}\int_0^1\int_0^{2\pi} ds\,d\chi\,\{b_1|I_6|^2 + b_2|I_7|^2 + b_3|I_8|^2\} \tag{6.18}$$

with

$$I_6 = \hat{X}\,,\quad I_7 = \hat{V} + \hat{Z} - \hat{\beta}_\chi\hat{X}\,,\quad I_8 = \hat{Z}\,,$$

$$b_1 = \frac{2\varrho\psi}{Js|\nabla\psi|^2}\,,\quad b_2 = \frac{\varrho J|\nabla\psi|^2}{2s\psi_p|\boldsymbol{u}|^2}\,,\quad b_3 = \frac{\varrho J T^2}{2s\psi_p|\boldsymbol{u}|^2}\,,$$

where $\hat{\beta}_\chi = 2s\psi_p\beta_\chi$. To calculate the vacuum energy W_v two methods are provided. The first method uses a Green's-function technique, by which the problem is reduced to a coupled set of integral equations on the boundaries of the vacuum region. Then the vacuum energy is obtained as a surface integral, depending on the normal component of the displacement ξ on the plasma surface. This method is described by *Merkel* (1982).

The second approach described in Chap. 5 for an axisymmetric geometry treats the vacuum contribution formally in the same way as the plasma potential energy (*Gruber* et al. 1981b). The eigenvalue problem can then be solved in the whole domain in one step.

As described in Sect. 5.4, a coordinate system $(\tilde{\psi}, \tilde{\chi})$ is introduced as well as a pseudomagnetic field

$$\tilde{\boldsymbol{B}} = \tilde{T}\boldsymbol{u} + \boldsymbol{u} \times \nabla\tilde{\psi} \tag{6.19}$$

which is tangent to the $\tilde{\psi}$ surfaces and satisfies $\nabla \cdot \tilde{\boldsymbol{B}} = 0$. The component $\tilde{T}$ is an arbitrary function of $\tilde{\psi}$ and $\tilde{\chi}$. Since $\nabla \times \tilde{\boldsymbol{B}} \neq 0$, this pseudomagnetic field is not the real vacuum field. By analogy with the plasma, a pseudodisplacement $\tilde{\boldsymbol{\xi}}$ is introduced such that

$$\boldsymbol{A} = \tilde{\boldsymbol{\xi}} \times \tilde{\boldsymbol{B}} + \nabla\Phi \ . \tag{6.20}$$

This representation is only possible if $\tilde{\boldsymbol{B}} \cdot \nabla \neq 0$ which can be fulfilled by an adequate choice of $\tilde{T}$. The pseudodisplacement $\tilde{\boldsymbol{\xi}}$ is expressed in terms of the two new dependent variables $\tilde{X}$ and $\tilde{V}$ by

$$\tilde{\boldsymbol{\xi}} = \frac{\tilde{X} T_{\mathrm{p}}}{|\nabla\tilde{\psi}|\, q_{\mathrm{p}}} \nabla\tilde{\psi} + \frac{\tilde{V}}{|\boldsymbol{u}|^2} \boldsymbol{u} \ , \tag{6.21}$$

where T_{p} and q_{p} are the values of $T(\psi)$ and $q(\psi)$ at the plasma side of Γ_{p}, respectively. The vector potential $\boldsymbol{A}$ then becomes

$$\boldsymbol{A} = -\tilde{V}\nabla\tilde{\psi} + \frac{\tilde{X} T_{\mathrm{p}}}{q_{\mathrm{p}}} \boldsymbol{u} - \frac{\tilde{X}\tilde{T} T_{\mathrm{p}}}{|\nabla\tilde{\psi}|^2 q_{\mathrm{p}}} (\boldsymbol{u} \times \nabla\tilde{\psi}) + \nabla\Phi \ . \tag{6.22}$$

The boundary condition at Γ_{p} (5.40) reads

$$\boldsymbol{n} \times \boldsymbol{A} = -\frac{\hat{X} T_{\mathrm{p}}}{q_{\mathrm{p}}} (T_{\mathrm{p}}\boldsymbol{u} + \boldsymbol{u} \times \nabla\psi) \quad \text{at} \quad \Gamma_{\mathrm{p}} \ , \tag{6.23}$$

where the quantities at the right-hand side are seen from the plasma side. These conditions are satisfied if $\tilde{X} = \hat{X}$ and

$$\tilde{T}(\psi, \chi) = T_{\mathrm{p}} |\nabla\tilde{\psi}| / |\nabla\psi| \equiv q_{\mathrm{p}} / (\tilde{J} |\boldsymbol{u}|^2) \quad \text{at} \quad \Gamma_{\mathrm{p}} \ . \tag{6.24}$$

The last identity results from the continuity of χ across the PVI, where $\tilde{J}^{-1} = \boldsymbol{u} \cdot (\nabla\tilde{\psi} \times \nabla\tilde{\chi})$ is the inverse of the Jacobian in the vacuum region. We extend the relation (6.24) to the full vacuum region by writing

$$\tilde{T}(\tilde{\psi}, \tilde{\chi}) = [q_{\mathrm{p}} / (\tilde{J} |\boldsymbol{u}|^2)]\, g(\tilde{\psi}) \tag{6.25}$$

with $g(\psi_{\mathrm{p}}) = 1$. This function $g(\tilde{\psi})$ is used to ensure that

$$\begin{gathered} \tilde{\boldsymbol{B}} \cdot \nabla = \frac{1}{\tilde{J}} \left[\frac{\partial}{\partial\tilde{\chi}} + \tilde{q} \frac{\partial}{\partial z} \right] = \frac{1}{\tilde{J}} \left[\frac{\partial}{\partial\tilde{\chi}} + \mathrm{i}k\tilde{q} \right] , \\ \tilde{q} = q_{\mathrm{p}} g(\tilde{\psi}) + \tilde{\alpha} \quad \text{and} \quad \tilde{\alpha} = \tilde{J} \nabla z \cdot (\boldsymbol{u} \times \nabla\tilde{\psi}) \end{gathered} \tag{6.26}$$

is invertible in the vacuum. By choosing $g(\tilde{\psi})$ such that

$$\oint_{\tilde{\psi}=\text{const}} \tilde{q}dl = \oint_{\psi_p=\text{const}} \tilde{q}_p dl = q_p + \oint_{\psi_p=\text{const}} \tilde{\alpha}_p dl \tag{6.27}$$

we ensure that $\tilde{\boldsymbol{B}}\cdot\boldsymbol{\nabla}$ is invertible everywhere if it is invertible at the plasma surface.

Introducing the variable s as in (5.60) and substituting all these expressions into the vacuum potential energy, (5.37) leads to

$$W_v = \frac{1}{2}\int_1^2\int_0^{2\pi} ds\, d\tilde{\chi} \left(\frac{T_p}{\tilde{T}}\right)^2 \{a_1|I_1|^2 + a_2|I_2|^2 + a_3|I_3|^2\} \ , \tag{6.28}$$

where the constants a_1, a_2, a_3 and the expressions I_1, I_2, and I_3 are identical to those in the plasma region, (6.16), except for the quantity H which is now given by

$$H = -\frac{1}{g}\frac{dg}{ds} = \frac{s\psi_p h}{\pi g q_p}\int_0^{2\pi} |\boldsymbol{u}|^4 \tilde{J}\, d\tilde{\chi} \ . \tag{6.29}$$

6.3 Applications

6.3.1 Straight Heliac

For the Heliac device – a stellarator with a helical axis and a kidney-shaped plasma cross section which is approximately helically symmetric – a high-beta stability limit is predicted because of its deep magnetic well and because of the short connection length between unfavorable and favorable curvature regions (*Furth* et al. 1966).

To study equilibrium and stability properties of such configurations, the helically symmetric analog of Heliac can serve as a suitable model. The results obtained for the straight equilibria will also become quantitatively relevant for a toroidal Heliac with a high number of helical periods, when the toroidal curvature is small compared with the helical curvature, and when the properties of the equilibria are dominated by the helical geometry.

To demonstrate ideal MHD stable equilibria, a sequence of equilibria is generated by varying (essentially) one parameter, the others being kept fixed. With an appropriately chosen parameter, one will obtain a sequence of equilibria which cross the marginal stability point and become stable for part of the parameter interval. For these parameter studies, the *Mercier* (1962) and the ballooning mode stability criteria (*Correa-Restrepo* 1978) evaluated on each magnetic surface, turn out to be very useful in searching for stable equilibria, because we find that the high-n-number ballooning criterion is sufficient for stability (*Gruber* et al. 1981d) for the types of equilibria considered here.

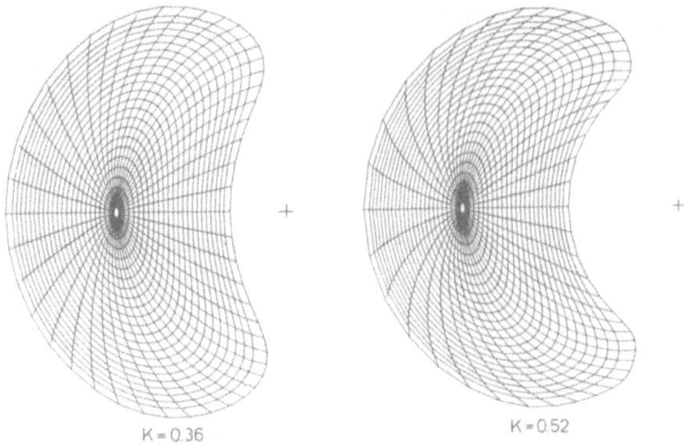

Fig. 6.4. Flux surface cross sections of straight Heliac equilibria in the plane $z=\text{const}$. A stable ($\kappa=0.52$) and an unstable ($\kappa=0.36$) case of a sequence are shown, characterized by the following set of parameters: $\langle\beta\rangle=0.21$, $E=2$, $S_{nn}=(-1)^n(n-1.8)/2.2$, $n=3,\ldots 10$; $f_0=1$, $f_1=-p_0$, $p_0=-1.54, -1.60, -1.62, -1.65, -1.68$, $p_1=-3.38, -3.245, -3.2, -3.11, -3.02$; $\kappa=0.52, 0.46, 0.44, 0.40, 0.36$. The helical aspect ratio is about $A=2.5$. From *Merkel* et al. (1983)

By varying the curvature κ of the helical axis between $\kappa=0.36$ and $\kappa=0.52$, a sequence of Heliac equilibria is generated. The average plasma-beta of $\langle\beta\rangle=0.21$ and the coefficients S_{nn} are kept fixed. In order to approximate the condition of vanishing net longitudinal current, it is sufficient to take profile functions linear in ψ with appropriately selected coefficients:

$$\begin{aligned} T(\psi) &= 1-p_0\psi \\ p'(\psi) &= p_0+p_1\psi \ . \end{aligned} \tag{6.30}$$

In Fig. 6.4, the kidney-shaped flux surfaces of the most unstable equilibrium ($\kappa=0.52$) are shown. The aspect ratio of the equilibria is about $A=1/ha\approx 2.5$, where a is the mean value of the plasma radius.

Figure 6.5 shows the profiles of the rotational transform ι versus $\sqrt{\psi}$, which is approximately proportional to the radius. The ι profiles are flat, with increasing shear at the plasma boundary. The corresponding pressure profiles are shown in Fig. 6.6. They vary only slightly within the equilibrium sequence.

Concerning the stability of this sequence, we find that the Mercier and ballooning-mode marginal point is reached for $\kappa\approx 0.48$, i.e., the equilibrium with $\kappa\approx 0.52$ is stable, while equilibria with lower values of κ are unstable. For this particular sequence, the stability limits of the Mercier and the ballooning-mode criteria practically coincide, a fact which is in general not true.

To complete the stability analysis, the HERA spectral code is applied to study the low-m-number modes. The eigenvalues ω^2 calculated by HERA depend on the wave number k, and may be characterized by the poloidal and radial node numbers of the eigenmodes. Only the lowest eigenvalues for fixed k, whose eigenmodes generally have no radial nodes, are calculated.

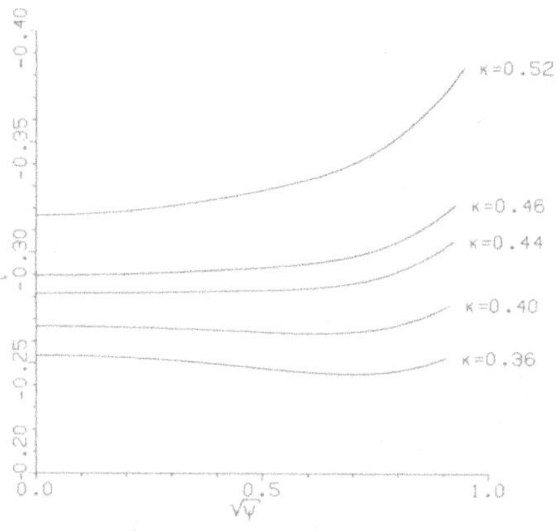

Fig. 6.5. Profiles of the rotational transform ι versus normalized flux function for the Heliac equilibrium sequence. From *Merkel* et al. (1983)

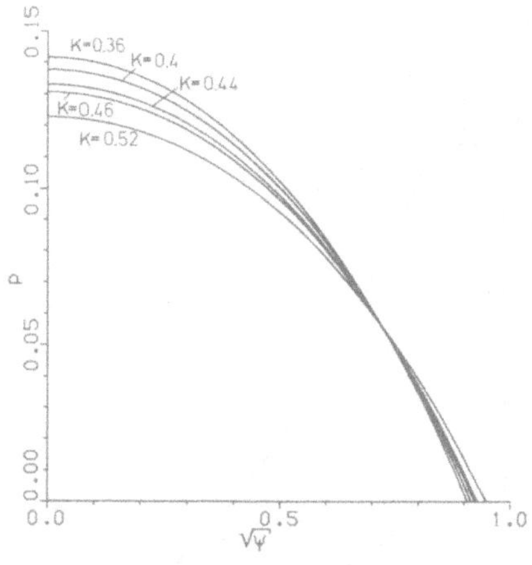

Fig. 6.6. Profiles of the pressure p versus normalized flux function for the Heliac equilibrium sequence. From *Merkel* et al. (1983)

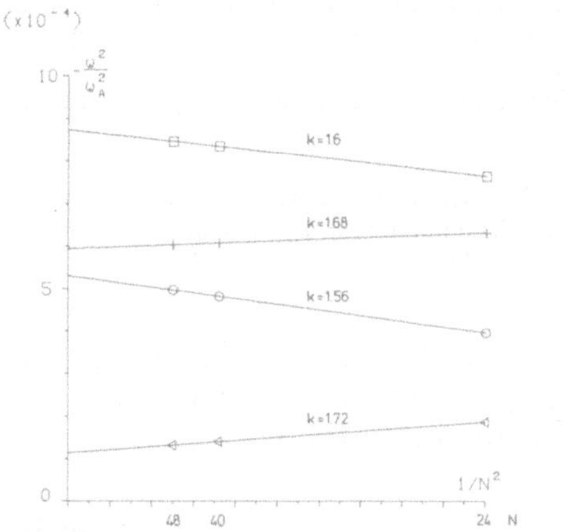

Fig. 6.7. Eigenvalues ω^2 of the internal $m=2$ mode of the Heliac equilibrium ($\kappa=0.36$) versus $1/N^2$, where $N=N_\psi=N_\chi$ is the number of radial and poloidal intervals

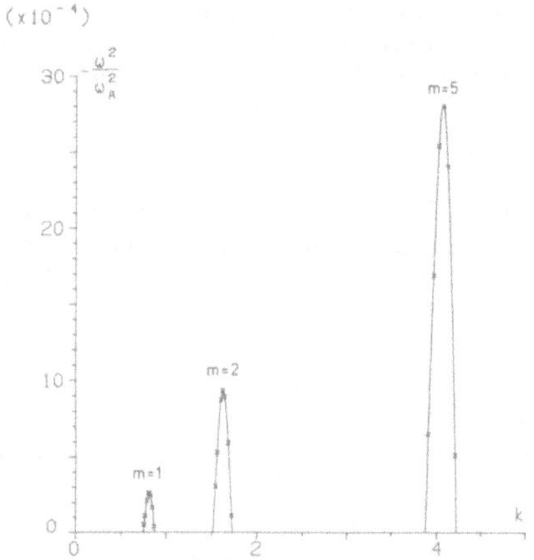

Fig. 6.8. The eigenvalues ω^2 of the internal $m=1, 2, 5$ modes versus k for the Heliac equilibrium with $\kappa=0.36$ (see Fig. 6.4)

To find the extrapolated values of the eigenvalues the numbers of radial and poloidal intervals are varied between $N=24$ to $N=48$. In all cases presented here, an equal number of radial (N_ψ) and poloidal (N_χ) intervals are chosen $N=N_\psi=N_\chi$. In Fig. 6.7 eigenvalues of an internal mode $m=2$ for a Heliac equilibrium with $\kappa=0.36$ (Fig. 6.4) are plotted versus $1/N^2$. The results show an excellent quadratic convergence for all the k values chosen.

For a given m, the eigenvalues ω^2 have maxima as a function of k. In Fig. 6.8 the eigenvalues ω^2 of the Heliac equilibrium ($\kappa=0.36$) are plotted as functions of

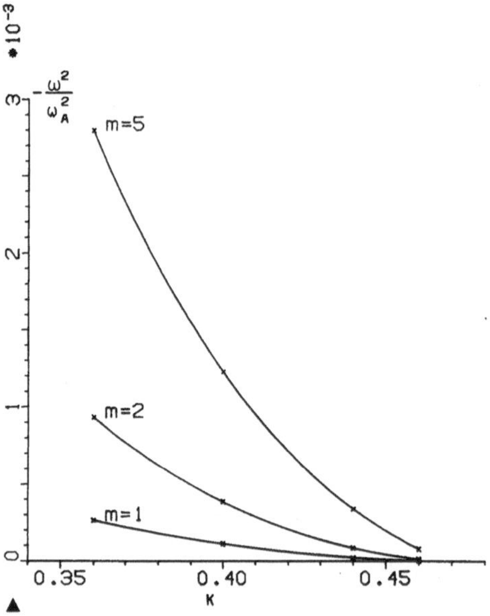

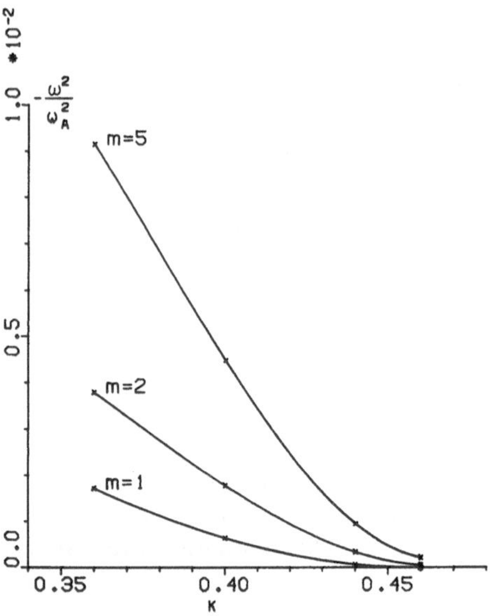

Fig. 6.9. Maxima of the eigenvalues for the $m=1, 2, 5$ internal modes versus curvature κ of the magnetic axis for the Heliac equilibrium sequence. The maxima at $\kappa=0.46$ are $|\omega^2/\omega_A^2|=7.8\times10^{-4}$, 1.5×10^{-5}, 3.8×10^{-6}, where $\omega_A^2=T^2(\psi)/4\pi\varrho(0)a_p^2$; $T(\psi)=$ profile function at the plasma surface ψ, $\varrho(0)=$ density at the magnetic axis, and $a_p=$ mean value of the plasma radius. From *Merkel* et al. (1983)

Fig. 6.10. Maxima of the eigenvalues for the $m=1, 2, 5$ external modes versus curvature κ of the magnetic axis for the Heliac equilibrium sequence. The maxima at $\kappa=0.46$ are $|\omega^2/\omega_A^2|=2.2\times10^{-4}$, 5.6×10^{-5}, 5.1×10^{-6}. From *Merkel* et al. (1983)

k. Shown are the modes $m=1$, $m=2$, and $m=5$. The maxima are typically achieved when the resonance condition $kq=m$ is satisfied in the plasma region, where the helical q factor is related to the rotational transform by

$$q=-\frac{1}{(\iota+1)h}=\frac{1}{2\pi}\oint\tilde{q}\,d\chi\ , \tag{6.31}$$

where the integration path is along a cut of a $z=$const with a $\psi=$const surface.

In all cases considered, the values of the maxima increase with m. These maxima (with respect to k) of the internal modes are shown for the sequence of equilibria in Fig. 6.9, i.e., in dependence of κ. For $\kappa=0.52$, unstable eigenvalues could not be found. The corresponding eigenvalues of the external modes are shown in Fig. 6.10. These external modes are calculated with a vacuum region bounded by a conducting shell of the same shape as the plasma boundary, similarly enlarged by a factor of $R_w=2$. This factor is chosen so large that the stabilizing influence of the wall on the external modes becomes unimportant. The eigenvalues are somewhat larger than the maxima of the internal modes, but the internal and external modes become marginal for the same value of κ. It

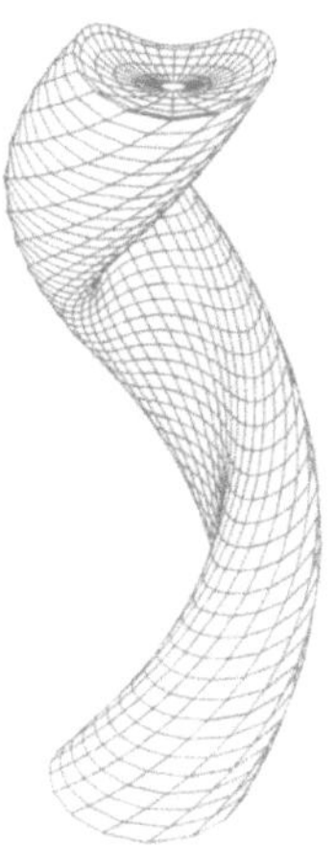

Fig. 6.11. MHD-stable Heliac configuration with $\langle\beta\rangle=0.29$. The parameters of the equilibrium are: $E=1.8$, $\kappa=0.6$, $S_{nn}=(-1)^n(n-1.8)/4$, $n=3,\ldots 10$; $f_0=1$, $f_1=-p_0$, $p_0=-2.04$, $p_1=-6.9$, $\iota_0=-0.32$. The aspect ratio is $A=2.24$. From *Merkel* et al. (1983)

appears that, at least for this particular sequence of longitudinal net-current-free equilibria, the stability properties are completely determined by the internal modes. The plots show a continuous transition to the stable region close to $\kappa=0.48$.

When sequences of equilibria are generated along these lines, stable equilibria with higher stability $\langle\beta\rangle$ limits are found. Increasing the curvature κ and choosing a smaller aspect ratio of $A=2.24$, a stable equilibrium with $\langle\beta\rangle=0.29$ is found. The flux surfaces and the parameters by which the equilibrium is determined are given in Fig. 6.11.

6.3.2 Straight Heliotron Equilibria

We consider now a second sequence of equilibria which can be characterized as the straight analogon of the Heliotron (*Uo* et al. 1981). The configuration, the flux surfaces and the parameters of the equilibrium sequence are given in Fig. 6.12. The magnetic axis is straight and the aspect ratio of a field period is $A=0.7$. The pressure profile is approximately $p=\beta_0(1-(\varrho/a)^4)$, so that unstable behavior at the center is avoided. The rotational transform per helical period ι varies from the axis to the edge from $\iota_0=-0.052$ to $\iota_E=-0.211$. The second arbitrary equilibrium function T and the pressure profile p' are adjusted in detail, so that the net longitudinal current vanishes, to a good approximation. Then the plasma boundary, the pressure profile and the rotational transform are independent of beta, to a good approximation, for the range of betas considered: $0\leq\beta\leq 0.06$. For this sequence of equilibria, the Mercier and ballooning stability limit occurs at $\langle\beta\rangle\approx 0.004$.

Results of the low-m-number mode analysis are shown in Fig. 6.13. The maximal growthrates of the m-number external modes are plotted as a function of $\langle\beta\rangle$. For the conducting wall, $R_w=2$ has been chosen. A very similar picture has been obtained (*Merkel* et al. 1983) when only internal modes are considered.

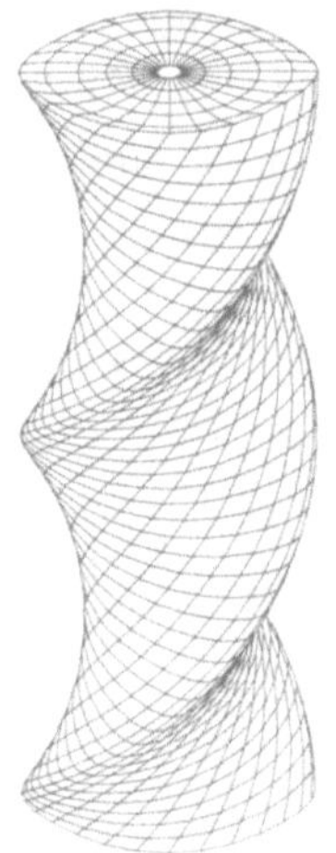

Fig. 6.12

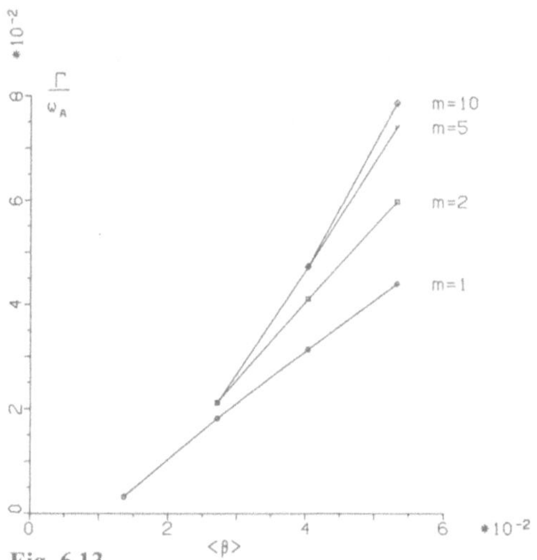

Fig. 6.13

Fig. 6.12. Straight Heliotron configuration. (From *Merkel* et al. 1983)

Fig. 6.13. Maxima of the growth rates for the $m=1$, 2, 5, 10 external modes versus $\langle\beta\rangle$. The parameters of the equilibrium sequence are: $\kappa=0$, $\tau=1$, $E=1.393$, $\langle\beta\rangle=0.053$, 0.040, 0.027, 0.014; $f_2=0.24$, 0.18, 0.12, 0.06; $p_1=-0.48$, -0.36, -0.24, -0.12; $p_2=0.06$, 0.045, 0.03, 0.015; $p_3=0.1$, 0.075, 0.05, 0.025; $S_{22}=(E-1/E)/2$; $S_{44}=-0.0116$, -0.0087, -0.0058, -0.0029. For the conducting shell bounding the vacuum region, the same shape was chosen as for the plasma boundary enlarged by a factor of $R_w=2$. (From *Merkel* et al. 1983)

This means that, as in the Heliac case, the stability properties are completely determined by the internal modes.

Formally extrapolating the curves of the growthrate, one obtains a marginal point for stability close to $\langle\beta\rangle=0.01$. (Because of the large shear of these equilibria, a detailed investigation of the structure of the modes in the gap between $0.004\leq\langle\beta\rangle\leq0.01$ is beyond the numerical resolution of HERA). These results are in contrast to the approximative treatment by means of stellarator expansion by *Wakatani* et al. (1979), which led to values of critical $\langle\beta\rangle\geq0.02$ with respect to low-*m*-number stability.

Convergence studies and comparisons between HERA and the three-dimensional stability code BETA (*Bauer* et al. 1978) for the straight HELIOTRON are presented in *Betancourt* et al. (1983).

6.3.3 Large-*k* Ballooning Modes

The HERA code allows a high k, m-number mode analysis by eliminating the fast poloidal variation of the modes with a suitable phase factor (*Gruber* et al. 1981b). This option of HERA is applied to an unstable equilibrium with a mixture of $l=1,2,3$ field content[1] and with approximately vanishing net

1 l is a number which characterizes the poloidal variation of the magnetic field. A mixture of $l=1,2,3$ for example can be obtained with 3 helical windings

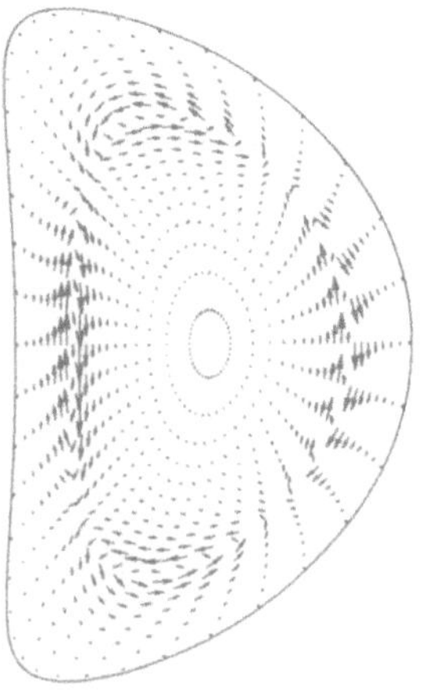

Fig. 6.14

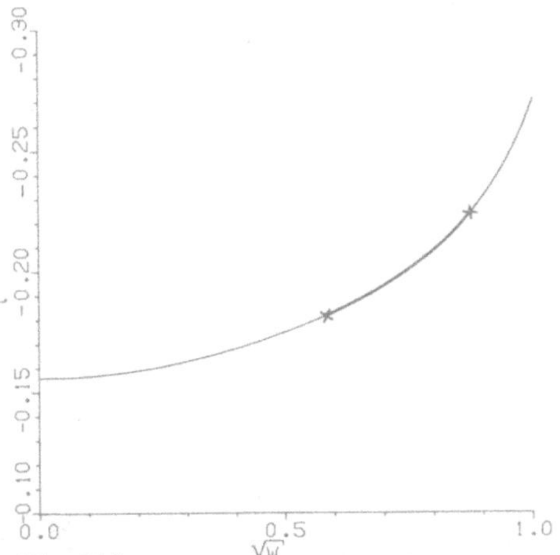

Fig. 6.15

Fig. 6.14. Flux surface cross section and eigenfunction of an $m=2$ mode of an equilibrium with a mixture of $l=1,2,3$ field content, with $\kappa=0.436$, $E=1.436$, $\langle\beta\rangle=0.05$, $p_1=15$, $p_2=61$, $p_3=-180$, $f_0=1$, $f_2=-9.5$, $S_{33}=-0.6$. (From *Merkel* et al. 1983)

Fig. 6.15. Profile of rotational transform versus $\sqrt{\psi}$ (ψ=normalized flux function) for the $l=1,2,3$ equilibrium. The heavy line indicates the Mercier unstable region. (From *Merkel* et al. 1983)

longitudinal current (*Gruber* et al. 1981d). The aspect ratio per helical period is about $A=3.5$ and the average beta is $\langle\beta\rangle=0.052$. The flux surface cross sections and the parameters determining the equilibrium are given in Fig. 6.14. In Fig. 6.15, the rotational transform ι is shown as a function of ψ. The shear of the equilibrium considerably increases towards the plasma boundary.

When the Mercier and the ballooning mode criteria on each magnetic surface are evaluated, the equilibrium is found to be stable around the magnetic axis and close to the plasma boundary, while it is unstable with respect to both criteria in the intervening region. The marginal points coincide for the Mercier and the ballooning mode criteria; they are marked in the ι plot in Fig. 6.15. In the range of $0.18 < -\iota < 0.22$, the equilibrium is unstable.

The eigenvalues of internal and external modes are calculated up to values of k which correspond to a mode number of $m=30$. In the case of external modes the plasma is surrounded by a conducting shell of the same shape as the plasma, similarly enlarged by a factor $R_w=2$.

In Figs. 6.16, 17, the eigenvalues of the internal mode $m=5$, and of the external modes $m=2$ are plotted versus k. As a general result, it can be stated that unstable modes practically only exist for values of k for which the resonant surface lies in the Mercier and ballooning unstable region. In Figs. 6.16, 17, the values of k are indicated for which the resonance condition is satisfied within the Mercier- and ballooning-mode unstable region. For the low-m-number external modes, this property does not hold as rigorously as for the internal modes (Fig. 6.17).

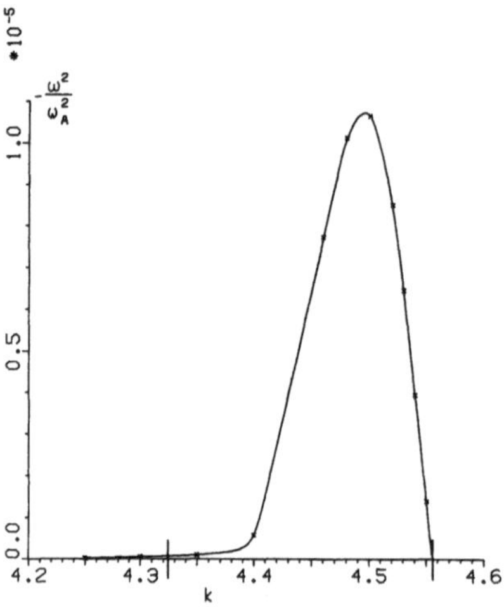

Fig. 6.16. Eigenvalues of the internal $m=5$ mode versus k for the $l=1, 2, 3$ equilibrium. For values of k with $4.32<k<4.56$, the resonant surface lies in the unstable region. (From *Merkel* et al. 1983)

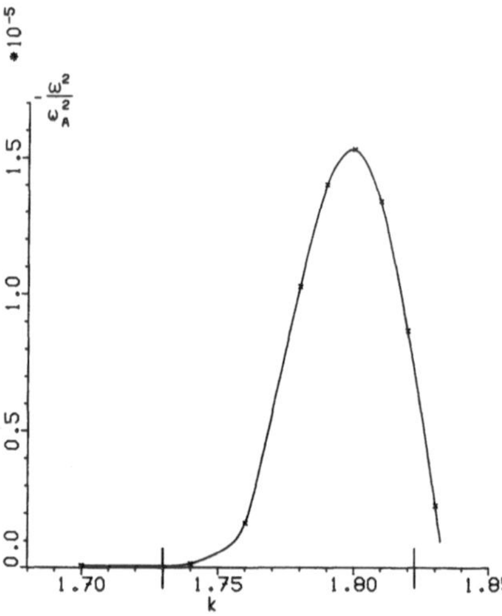

Fig. 6.17. Eigenvalues of the external $m=2$ mode versus k for the $l=1, 2, 3$ equilibrium ($R_w=2$). For values of k with $1.73<k<1.82$, the resonant surface lies in the unstable region. (From *Merkel* et al. 1983)

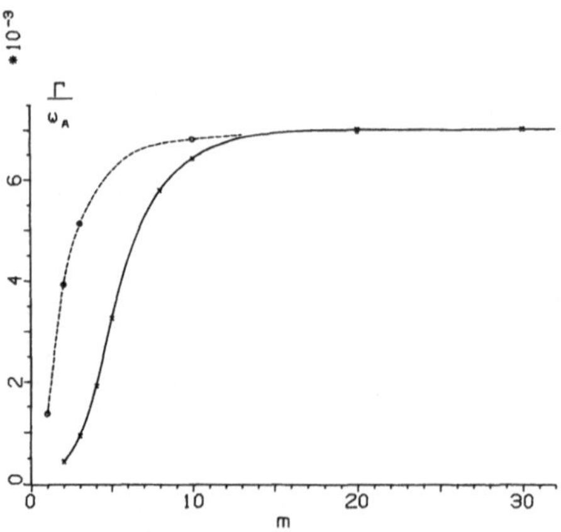

Fig. 6.18. Growth rates of internal (×) and external (0) modes versus the lumber m of poloidal nodes for the $l=1, 2, 3$ equilibrium

The eigenfunctions of all these modes are very localized around the resonant surface. This may be due mainly to the shear of the magnetic field, but apparently the eigenfunctions become small in the Mercier and ballooning stable region. In Fig. 6.14, the eigenfunction of an external $m=2$ mode is shown.

The stability behavior of the equilibrium appears to be completely determined by the internal modes because all eigenfunctions are restricted to the Mercier and ballooning mode unstable region, which leads to a natural interpretation of the Mercier criterion. The eigenvalues of the external modes are only slightly larger than those of the internal modes.

The results are summarized in Fig. 6.18, where the maxima of the growthrates are plotted versus the mode number m. The growthrates of the internal and external modes increase with m to the same finite limiting value for $m \to \infty$.

6.3.4 Conclusion

Helically symmetric equilibria without net longitudinal current and their linear MHD stability are studied. MHD-stable straight Heliotron equilibria up to $\langle\beta\rangle = 0.01$ and stable straight Heliac equilibria up to $\langle\beta\rangle \approx 0.3$ are obtained. Applying the large-k-number option of the HERA code, it is demonstrated that also high-m-number ballooning modes can be studied numerically. The results are in good agreement with the ballooning mode criterion.

Although stability cannot be proved by the numerical solution of the eigenvalue problem, the lower limits for the unstable eigenvalues can be estimated to be $\omega^2/\omega_A^2 > -10^{-7}$ (corresponding to upper limits on growthrates of $|\omega/\omega_A| < 10^{-3}$). All eigenvalues are converged values, i.e., extrapolated to zero mesh-size. The growthrates shown are solutions of the linear stability operator. Whether these results have to be taken seriously depends on the influence of toroidicity, the non-linear behavior, such as the saturation of the modes, or non-ideal effects, such as the resistive ballooning modes. To tackle the non-linear behavior, non-linear 3D codes become more and more powerful. Since first comparative calculations (*Betancourt* et al. 1983) with the 3D BETA code (*Bauer* et al. 1978) and the HERA code showed encouraging agreement, it will soon be possible to reliably obtain the non-linear saturation behavior of pressure-driven modes. As to the resistive ballooning modes, their influence on the stability of equilibria with vanishing net longitudinal current (stellarators) is also becoming tractable (*Correa-Restrepo* 1982).

7. Similar Problems

7.1 Similar Problems in Plasma Physics

7.1.1 Resistive Spectrum in a Cylinder

The full non-linear resistive MHD equations are those given in (2.1, 2) in which Ohm's law is replaced by

$$\frac{\partial \boldsymbol{B}}{\partial t}=\nabla\times(\boldsymbol{v}\times\boldsymbol{B})-\frac{1}{s}\nabla\times(\eta_0\nabla\times\boldsymbol{B}) \ , \tag{7.1}$$

where s is the magnetic Reynolds number and η_0 is the resistivity tensor. It is clear that it is not further possible to time-integrate (7.1) which forces us to use the velocity $\boldsymbol{v}$ as an unknown, and not the displacement $\boldsymbol{\xi}$. In addition, it is not possible to eliminate the pressure and the magnetic field in the linearized equations. As a consequence, we have to solve the system of (3.4) to which the term with resistivity given in (7.1) has to be added. Using $\nabla\cdot\boldsymbol{B}=0$ to eliminate the poloidal component of the perturbed magnetic field, and demanding that resistivity be constant during the time we are interested in, we obtain the following set of linearized equations:

$$\begin{aligned}
i\omega\varrho v_1 &= -\frac{r}{m}\left[\frac{m}{r}\tilde{p}+B_\theta b_1'-\left(\frac{m}{r}B_z-kB_\theta\right)b_3\right]'+Fb_1-\frac{2}{m}B_\theta b_1'+\frac{2k}{m}B_\theta b_3\\
i\omega\varrho v_2 &= \frac{m}{r^2}\tilde{p}+\frac{1}{r^2}(B_\theta r)'b_1-\frac{k}{m}B_z b_1'+\frac{K^2}{m}B_z b_3\\
i\omega\varrho v_3 &= k\tilde{p}-\frac{rK^2}{m}B_\theta b_3+\frac{krB_\theta}{m}b_1'\\
i\omega\tilde{p} &= -p_0'v_1-\gamma p_0(v_1'+mv_2+kv_3)\\
i\omega b_1 &= Fv_1+\frac{\eta_0}{s}\left[b_1''+\frac{b_1'}{r}-K^2b_1-\frac{2k}{r}b_3\right]\\
i\omega b_3 &= Fv_3-B_z(v_1'+mv_2+kv_3)+\frac{\eta_0}{s}\left[\left(b_3'-\frac{b_3}{r}\right)'-K^2b_3\right]\\
&\quad -\frac{\eta_0'}{s}\left(\frac{kb_1}{r}-\left(\frac{b_3}{r}\right)'\right) ,
\end{aligned} \tag{7.2}$$

where

$$F = kB_z + \frac{m}{r} B_\theta$$
$$K^2 = k^2 + \frac{m^2}{r^2} \ , \tag{7.3}$$

and ϱ, B_θ, B_z, p_0, and η_0 are the static equilibrium quantities. As unknowns we have chosen

$$\begin{aligned} v_1 &= rv_r \qquad \tilde{p} = r\delta p \\ v_2 &= \mathrm{i}v_\theta \qquad b_1 = -\mathrm{i}r\delta B_r \\ v_3 &= \mathrm{i}rv_z \qquad b_3 = r\delta B_z \ . \end{aligned} \tag{7.4}$$

The expansion of all the unknowns in the weak form of (7.2) in terms of hat-function elements leads to a polluted approach.

No pollution is seen when we choose Hermite-cubic elements for v_1 and b_1 and Lagrange elements of second order for the other components (*Kerner* et al. 1984). Due to the inclusion of resistivity, the very localized and the singular modes become broad and global. This is the reason why cubic elements have been chosen. As we have seen in Chap. 3 they give a better representation of the global modes.

7.1.2 Non-Linear Plasma Wave Equation (with M.C. Festeau-Barrioz)

Similar numerical problems to those discussed in Chap. 1, 3, and 4 arise when solving the non-linear eigenvalue problem (*Festeau-Barrioz* and *Weibel* 1982)

$$\nabla \times (\nabla \times \boldsymbol{E}) = -\omega^2 \varepsilon(\boldsymbol{E}) \boldsymbol{E} \tag{7.5}$$

in a conducting cylinder of radius a and height b. Here, $\boldsymbol{E}$ is the electrostatic field which is written

$$\boldsymbol{E} = (E_r(r,z), E_\theta(r,z), E_z(r,z))\, \mathrm{e}^{im\theta - \mathrm{i}\omega t} \tag{7.6}$$

and $\varepsilon(\boldsymbol{E})$ is the dielectric tensor including the non-linearity.

In a particular case, $\varepsilon = I$, the identity matrix, and $\omega^2 = 0$ is an infinitely degenerate eigenvalue corresponding to an eigenvector exactly satisfying $\nabla \times \boldsymbol{E} = 0$. In this linear case, the variational form is written simply

$$\delta \int_\Omega [(\nabla \times \boldsymbol{E})^2 - \omega^2 E^2] dx = 0 \ , \tag{7.7}$$

restricted by

$$\boldsymbol{E} \times \boldsymbol{n} = 0 \quad \text{on} \quad \delta\Omega \ . \tag{7.8}$$

Reminding that

$$\nabla \times \boldsymbol{E} = \left[\frac{1}{r}\left(imE_z - \frac{\partial rE_\theta}{\partial z}\right), \frac{\partial E_r}{\partial z} - \frac{\partial E_z}{\partial r}, \frac{1}{r}\left(\frac{\partial rE_\theta}{\partial r} - imE_r\right)\right] \tag{7.9}$$

we propose to choose as variables the covariant components (E_r, rE_θ, E_z).

Spectrum pollution is obtained when we expand in terms of basis functions, linear in each variable, all three vector components of $\boldsymbol{E}$. An unpolluted approach is obtained when we choose these elements only for the component rE_θ. At the same time we have to take for E_r basis functions which are linear in z, but discontinuous in r, and for E_z basis functions which are linear in r but discontinuous in z. It is easy to show that with this choice of elements we are consistent with the boundary conditions (7.8) and it is possible to satisfy identically $\nabla \times \boldsymbol{E} = 0$ and $\omega = 0$.

7.1.3 Alfvén and ICRF Heating in a Tokamak (with K. Appert, T. Hellsten, J. Vaclavik, and L. Villard)

In radiofrequency heating experiments resonant effects are used to couple the energy launched by the antenna to a movement inside the plasma. In the low-frequency Alfvén wave heating case (frequency up to a few MHz) the resonant surface is a constant pressure or constant flux surface. In a multi-species plasma, increasing the frequency towards $\omega = \omega_{\mathrm{ci}}$, where ω_{ci} is the ion cyclotron frequency of the order of a few tenths of MHz, a second resonance surface, the so-called ion-ion hybrid resonance, enters the plasma.

To describe the coupling between the antenna and the plasma we use the cold plasma approximation, neglecting effects of the order of m_e/m_i, the ratio of electron and ion mass, leading to $\boldsymbol{E} \cdot \boldsymbol{B} = 0$. In addition, we consider low-β plasmas, well described by force-free equilibria for which

$$\boldsymbol{J} = \alpha \boldsymbol{B} \ . \tag{7.10}$$

This model leads to the following wave equation (*Stix* 1962)

$$\nabla \times (\nabla \times \boldsymbol{E}) - \alpha \nabla \times \boldsymbol{E} = \begin{pmatrix} \varepsilon & ig \\ -ig & \varepsilon \end{pmatrix} \boldsymbol{E} \ , \tag{7.11}$$

where

$$\begin{aligned} \varepsilon &= \frac{\omega^2}{c_A^2} \sum_{\text{ions}} \frac{f_i}{1 - \omega^2/\omega_{\mathrm{ci}}^2} \\ g &= \frac{\omega^2}{c_A^2} \sum_{\text{ions}} \frac{f_i \omega/\omega_{\mathrm{ci}}}{1 - \omega^2/\omega_{\mathrm{ci}}^2} \end{aligned} \tag{7.12}$$

and

$$f_i = \frac{n_i m_i}{\sum\limits_{\text{ions}} n_j m_j} , \tag{7.13}$$

where n_i is the ion density of the species i. The torus is surrounded by a vacuum region which is isolated by an infinitely conducting wall as in the case described in Chap. 5. The antenna which launches a wave of the form $\exp(i\omega t + in\phi)$ is mounted in the vacuum region. The boundary conditions at the plasma-vacuum interface Γ_p are that the magnetic field pressure and the tangential component of the electric field be continuous. On the antenna we prescribe a jump of the magnetic field pressure, whereas at the conducting wall the tangential component of $\boldsymbol{E}$ vanishes. Introducing a variable transformation

$$\boldsymbol{E} = \boldsymbol{\xi} \times \boldsymbol{B} , \tag{7.14}$$

which automatically drops out the parallel component of $\boldsymbol{E}$, the continuity condition at Γ_p is

$$(\boldsymbol{n} \cdot \boldsymbol{\xi})\boldsymbol{B} + \boldsymbol{n} \times \boldsymbol{A} = 0 \quad \text{at} \quad \Gamma_p , \tag{7.15}$$

where $\boldsymbol{A}$ is the vector potential in the vacuum region and $\boldsymbol{n}$ is the normal vector to the plasma surface Γ_p. At the conducting wall

$$\boldsymbol{n} \cdot \boldsymbol{\xi} = 0 \quad \text{at} \quad \Gamma_v . \tag{7.16}$$

In the very-low-frequency Alfvén heating ($\omega \ll \omega_{ci}$), $g = 0$ and $\varepsilon = \omega^2$, where ω^2 is a normalized mean frequency for all the species. In this case we obtain the pressureless ideal MHD equations, in which one forces ω^2 and obtains a system of linear equations with a right-hand side, given by the jump of the magnetic pressure across the antenna. This case leads to the same variational form as the variational form for a torus presented in Chap. 5, when forcing the pressure and the variable Z to zero. We know that we are well able to describe heating when we are well able to resolve numerically the resonant surfaces. This is possible when choosing the finite hybrid element approach described in Chap. 4 to discretize the variational form of (7.11).

7.2 Similar Problems in Other Domains

7.2.1 Stability of a Compressible Gas in a Rotating Cylinder

Let us consider an infinitely long cylinder with radius $r = a$ which rotates around its axis with an angular velocity ω. This cylinder is filled with a compressible gas. The equations describing this situation are (*Chandrasekhar* 1961)

$$\frac{\partial \varrho}{\partial t}+\nabla \cdot(\varrho \boldsymbol{v})=0 \quad \text{(continuity)} \tag{7.17}$$

$$\frac{\partial(\varrho \boldsymbol{v})}{\partial t}+(\boldsymbol{v}\cdot\nabla)\boldsymbol{v}=-\nabla p+\mu\Delta\boldsymbol{v}+\frac{\mu}{3}\nabla(\nabla\cdot\boldsymbol{v}) \quad \text{(motion)} \tag{7.18}$$

$$\frac{\partial}{\partial t}\left(\varrho\frac{\boldsymbol{v}^2}{2}+\varrho\varepsilon\right)=-\nabla\cdot\left[\varrho\boldsymbol{v}\left(\frac{v^2}{2}+\frac{p}{\varrho}+\varepsilon\right)-v\sigma'-\chi\nabla T\right] \quad \text{(energy)} \tag{7.19}$$

$$p=\varrho\frac{RT}{M}\ , \quad \text{(state)} \tag{7.20}$$

where ϱ, $\boldsymbol{v}$, p, μ, ε, σ', χ, T, and R/M are the mass density, velocity, gas pressure, viscosity, internal energy, tensor for the constraints, thermal conductivity, gas temperature and a constant, respectively.

The system of equations permits a stationary solution ($\partial/\partial t=0$) of the form

$$\begin{aligned}\boldsymbol{u}_0&=(v_r, v_\theta, v_z, T, \varrho, p)_0\\&=(0, \Omega r, w_0(r), T_0, \varrho_0\,\mathrm{e}^{s^2r^2}, p_0\,\mathrm{e}^{s^2r^2})\ ,\end{aligned} \tag{7.21}$$

where s is a constant determined by the type of gas.

To study the stability of such a system, we perturb the equilibrium solution, writing the total vector $\boldsymbol{U}$ as

$$\boldsymbol{U}=\boldsymbol{u}_0+\boldsymbol{u}\ , \tag{7.22}$$

where $\boldsymbol{u}$ is an small perturbation of $\boldsymbol{u}_0$. It is possible to eliminate the pressure p in (7.18, 19) using (7.20). The perturbation vector $\boldsymbol{u}$ then includes five components:

$$\boldsymbol{u}=(\delta v_r, \delta v_\theta, \delta v_z, \delta T, \delta\varrho)=(\boldsymbol{u}_1, u_2)\ , \tag{7.23}$$

where $\boldsymbol{u}_1$ contains the first four components and u_2 the last one $\delta\varrho$. The perturbed quantities obey the boundary conditions

$$\boldsymbol{u}_1=0 \quad \text{for} \quad r=a\ . \tag{7.24}$$

If (r, θ, z) is a cylindrical coordinate system, a normal-mode analysis is performed on $\boldsymbol{u}$:

$$\boldsymbol{u}(r,\theta,z,t)=\sum_{m=-\infty}^{+\infty}\int_{-\infty}^{+\infty}\boldsymbol{u}_{mk}(r)\exp(ikz+im\theta+\Gamma_{mk}t)\,dk\ . \tag{7.25}$$

Replacing (7.22 and 25) in the fluid equations, and neglecting higher-order terms in the perturbation, one can show that it is possible to solve for given integer m and

$$
A = \begin{bmatrix}
-\frac{4\mu}{3\rho_0} & 0 & 0 & 0 & 0 \\
0 & -\frac{\mu}{\rho_0} & 0 & 0 & 0 \\
0 & 0 & -\frac{\mu}{\rho_0} & 0 & 0 \\
0 & 0 & 0 & -\frac{\chi}{C_\nu \rho_0} & 0 \\
0 & 0 & 0 & 0 & 0
\end{bmatrix}
$$

$$
B = \begin{bmatrix}
-\frac{4\mu}{3r\rho_0} & -\frac{im\mu}{3r\rho_0} & -\frac{i\mu k}{3\rho_0} & \frac{R}{M} & \frac{RT_0}{M} \\
-\frac{im\mu}{3r\rho_0} & -\frac{\mu}{r\rho_0} & 0 & 0 & 0 \\
-\frac{i\mu k}{3\rho_0} & 0 & -\frac{\mu}{r\rho_0} & 0 & 0 \\
\frac{p_0}{C_\nu \rho_0} & 0 & -\frac{2\mu W_0'}{C_\nu \rho_0} & -\frac{\chi}{C_\nu \rho_0 r} & 0 \\
1 & 0 & 0 & 0 & 0
\end{bmatrix} .
$$

$$
C = \begin{bmatrix}
ikW_0 + \frac{\mu}{\rho_0}\left(\frac{4}{3r^2} + \frac{m^2}{r^2} + k^2\right) & 2\omega + \frac{\mu}{\rho_0}\frac{7im}{3r^2} & 0 & \frac{R}{M}\frac{\rho_0'}{\rho_0} & -\omega^2 r + \frac{R}{M} T_0 \frac{\rho_0'}{\rho_0} \\
2\omega - \frac{\mu}{\rho_0}\frac{7im}{3r^2} & ikW_0 + \frac{\mu}{\rho_0}\left(k^2 + \frac{1}{r^2} + \frac{4m^2}{3r^2}\right) & \frac{\mu m k}{3r\rho_0} & \frac{im}{r}\frac{R}{M} & \frac{im}{r}\frac{R}{M} T_0 \\
\frac{dW_0}{dr} - \frac{\mu i k}{3\rho_0} & \frac{m\mu k}{3r\rho_0} & ikW_0 + \frac{\mu}{\rho_0}\left(\frac{4}{3}k^2 + \frac{m^2}{r^2}\right) & ik\frac{R}{M} & ik\frac{R}{M} T_0 \\
\frac{p_0}{rC_\nu \rho_0} - \frac{2ik\mu}{C_\nu \rho_0}\frac{dW_0}{dr} & \frac{im\, p_0}{rC_\nu \rho_0} & \frac{ikp_0}{C_\nu \rho_0} & ikW_0 + \frac{\chi}{C_\nu \rho_0}\left(k^2 + \frac{m^2}{r^2}\right) & 0 \\
\frac{1}{r} + \frac{\rho_0'}{\rho_0} & \frac{im}{r} & ik & 0 & ikW_0
\end{bmatrix}
$$

Fig. 7.1. Matrices A, B, and C of the stability problem of a compressible gas in a rotating cylinder

any k, a one-dimensional linear eigenvalue problem with the eigensolution $\boldsymbol{u}_{mk}(r)$ corresponding to the eigenvalue Γ_{km}. The system of differential equations is

$$
A\boldsymbol{u}'' + B\boldsymbol{u}' + C\boldsymbol{u} = \lambda \boldsymbol{u} \ , \tag{7.26}
$$

where $' = d/dr$ and $\lambda = -\Gamma_{mk} - im\omega$. The matrices A, B, and C have the form as shown in Fig. 7.1. One immediately realizes that the fifth line and the fifth column of A are zero. This implies that we have a differential system of mixed order (*Grubb* and *Geymonat* 1977). The system is of second order in the four components of $\boldsymbol{u}_1$, and of first order in the fifth component u_2. This is a similar problem to that considered in Sect. 1.2, where we had two components, one of second and one of first order.

We consider a weak formulation of Problem (7.26) which is of the form

"Find complex numbers λ and non-trivial $\boldsymbol{u}$ in space V satisfying
$a(\boldsymbol{u}, \boldsymbol{v}) = \lambda b(\boldsymbol{u}, \boldsymbol{v})$, for all $\boldsymbol{v} \in V$," (7.27)

where V is the set of all complex-valued functions given by $V=(H_0^1(0,a))^4 \times L^2(0,a)$; $a(\cdot,\cdot)$ and $b(\cdot,\cdot)$ are two continuous sesquilinear forms on V. Problem (7.27) is of the form (1.54) with the difference that $a(\cdot,\cdot)$ is not Hermitian (for more details, see *Lefèvre* and *Rappaz* 1978).

If we choose linear elements for all five vector components we obtain a polluted spectrum. If we perform partial integrations such that no derivative on u_2 and v_2 appears in the variational form, we can choose linear elements for the four components $\boldsymbol{u}_1$ and piecewise constant elements for u_2, and no pollution occurs, as can be proved mathematically (*Lefèvre* and *Rappaz* 1978).

7.2.2 Normal Modes in the Oceans

We consider a rotating, infinitely deep basin full of an incompressible fluid. The rotation speed is so slow that we can admit a static equilibrium state. Perturbing this state leads to the following set of differential equations:

$$\begin{aligned} \frac{\partial \xi}{\partial t} &= -\nabla \cdot \boldsymbol{u} \\ \frac{\partial \boldsymbol{u}}{\partial t} &= -\nabla \xi - f\boldsymbol{u} - \omega R\boldsymbol{u} \ , \end{aligned} \tag{7.28}$$

where $\partial/\partial t$ is the derivative with respect to the time, $\boldsymbol{u}=(u_1,u_2)$ is the horizontal volume transport, ξ is the height of the fluid above equilibrium level, R is the linear operator $R\boldsymbol{u}=(-u_2,u_1)$, $f\geq 0$ and ω are real constants representing friction and Coriolis terms.

The horizontal surface of the basin is denoted by Ω and the boundary by Γ; $\boldsymbol{n}$ is the exterior normal to Γ. Hence, $\boldsymbol{u}$ and ξ are functions of $(x,y)\in\Omega$. The existence of rigid walls implies that

$$\boldsymbol{u}\cdot\boldsymbol{n}=0 \quad \text{on} \quad \Gamma \ , \tag{7.29}$$

and the incompressibility condition gives

$$\int_\Omega \xi\, dx\, dy = 0 \ . \tag{7.30}$$

In order to determine the normal modes of the rotating basin, we are concerned with the eigenvalues $\lambda\in\mathbb{C}$ of the system:

$$\begin{aligned} &-\nabla\cdot\boldsymbol{u}=\lambda\xi \quad \text{in} \quad \Omega \ , \\ &-\nabla\xi - f\boldsymbol{u} - \omega R\boldsymbol{u} = \lambda\boldsymbol{u} \quad \text{in} \quad \Omega \ , \\ &\boldsymbol{u}\cdot\boldsymbol{n}=0 \quad \text{on} \quad \Gamma \ , \\ &\int_\Omega \xi\, dx\, dy = 0 \ . \end{aligned} \tag{7.31}$$

Notice that if $f=\omega=0$, $\lambda=0$ is an eigenvalue of infinite multiplicity, the corresponding eigenvectors are $\xi=0$, and $\boldsymbol{u}$ is such that $\nabla\cdot\boldsymbol{u}=0$ in Ω and $\boldsymbol{u}\cdot\boldsymbol{n}=0$ on Γ. In fact it is possible to show that for any f and ω, the operator given by (7.31) has a non-compact resolvent.

In order to give a weak formulation of (7.31) we define the space

$$H^1_*(\Omega)=\left\{w\in H^1(\Omega):\int_\Omega w\,dx\,dy=0\right\}; \tag{7.32}$$

moreover we use the following notations:

$$\langle v,w\rangle=\int_\Omega vw\,dx\,dy\,,\quad \text{if}\quad v,w\in L^2(\Omega)$$

$$\langle \boldsymbol{v},\boldsymbol{w}\rangle=\langle v_1,w_1\rangle+\langle v_2,w_2\rangle\,,$$

$$\text{if}\ \ \boldsymbol{v}=(v_1,v_2)\ \text{and}\ \boldsymbol{w}=(w_1,w_2)\in(L^2(\Omega))^2\,. \tag{7.33}$$

If X is the space $H^1_*(\Omega)\times L^2(\Omega)^2$, we introduce the continuous sesquilinear form a on X, defined by

$$a((\zeta,\boldsymbol{u}),(\xi,\boldsymbol{v}))=\langle\boldsymbol{u},\nabla\xi\rangle-\langle\nabla\zeta,\boldsymbol{v}\rangle-\langle f\boldsymbol{u}+\omega R\boldsymbol{u},\boldsymbol{v}\rangle \tag{7.34}$$

for $(\zeta,\boldsymbol{u})$ and $(\xi,\boldsymbol{v})\in X$.

A weak formulation of (7.31) is:

"Find $(\zeta,\boldsymbol{u})\in X$, $(\zeta,\boldsymbol{u})\neq 0$ and $\lambda\in\mathbb{C}$ such that
$a((\zeta,\boldsymbol{u}),(\xi,\boldsymbol{v}))=\lambda(\langle\zeta,\xi\rangle+\langle\boldsymbol{u},\boldsymbol{v}\rangle)$
for all $(\xi,\boldsymbol{v})\in X$." (7.35)

If we subdivide the domain Ω by triangles and choose as basis functions continuous piecewise polynomials of degree 1 for ζ with zero integral over the domain Ω and piecewise constant elements for the two components of $\boldsymbol{u}$, no pollution appears. This has been shown by *Descloux* et al. (1981).

Appendices

A: Variational Formulation of the Ballooning Mode Criterion

For the ballooning mode criterion we compute the stability index τ [defined in (2.31, 32)] for modes with toroidal wave numbers of $n=\infty$. What we have to do is to check the sign of the plasma potential energy W_p (5.52–54). We first minimize with respect to the toroidal component Z, which implies setting the fourth quadratic term in (5.52) to zero. The two remaining unknowns X and V are expanded around the marginal solution, and for high values of n:

$$\begin{aligned} \hat{X} &= \left(X_0+\frac{1}{n}X_1+\dots\right)\exp[-inq(\chi-\chi_0)] \\ \hat{V} &= \left(V_0+\frac{1}{n}V_1+\dots\right)\exp[-inq(\chi-\chi_0)] \end{aligned} \tag{A.1}$$

The different quadratic terms then are written

$$\begin{aligned} I_1 &= \frac{\partial X_0}{\partial \chi}+\mathrm{O}\left(\frac{1}{n}\right) \\ I_2 &= D=\frac{\partial X_0}{\partial s}-\mathrm{i}n(\chi-\chi_0)\frac{dq}{ds}\left(X_0+\frac{1}{n}X_1\right) \\ &\quad +\frac{\partial V_0}{\partial \chi}-inq\left(V_0+\frac{1}{n}V_1\right)+\mathrm{O}\left(\frac{1}{n}\right) \\ I_3 &= D+HX_0+\hat{\beta}_\chi\frac{\partial X_0}{\partial \chi}-\frac{\partial V_0}{\partial \chi}+\mathrm{O}\left(\frac{1}{n}\right) \\ I_5 &= X_0+\mathrm{O}\left(\frac{1}{n}\right) , \end{aligned} \tag{A.2}$$

and the integration in χ has to be extended from $-\infty$ to $+\infty$ (*Connor* et al. 1978). In highest order the coefficient multiplying n must vanish:

$$(\chi-\chi_0)\frac{dq}{ds}X_0+qV_0=0 \ . \tag{A.3}$$

This equation (A.3) will be used to eliminate V_0. Introducing the new variable

$$U_0 = D \ , \tag{A.4}$$

which includes the variables X_1 and V_1, and after elimination of V_0 in I_3 one obtains

$$W_p(n=\infty) = \frac{1}{2}\int_0^1 \int_{-\infty}^{+\infty} ds\, d\chi \left\{ a_1 \left| \frac{\partial X_0}{\partial \chi} \right|^2 + a_2 |U_0|^2 \right.$$
$$\left. + a_3 \left| U_0 + H X_0 + \hat{\beta}_\chi \frac{\partial X_0}{\partial \chi} + \frac{dq}{ds} X_0/q + (\chi-\chi_0)\frac{dq}{ds}\frac{\partial X_0}{\partial \chi} \Big/ q \right|^2 - a_5 |X_0|^2 \right\} . \tag{A.5}$$

Combining the second and the third quadratic terms leads to the equivalent expression ($' = d/ds$).

$$W_p(n=\infty) = \frac{1}{2}\int_0^1 \int_{-\infty}^{+\infty} ds\, d\chi \left\{ a_1 \left| \frac{\partial X_0}{\partial \chi} \right|^2 \right.$$
$$+ \frac{a_2 a_3}{a_2 + a_3} \left| H X_0 + \hat{\beta}_\chi \frac{\partial X_0}{\partial \chi} + \frac{q'}{q} X_0 + (\chi-\chi_0)\frac{q'}{q}\frac{\partial X_0}{\partial \chi} \right|^2$$
$$+ (a_2 + a_3) \left| U_0 + \frac{a_3}{a_2 + a_3}\left(H X_0 + \hat{\beta}_\chi \frac{\partial X_0}{\partial \chi} + \frac{q'}{q} X_0 + (\chi-\chi_0)\frac{q'}{q}\frac{\partial X_0}{\partial \chi} \right) \right|^2$$
$$\left. - a_5 |X_0|^2 \right\} . \tag{A.6}$$

Variation with respect to U_0 gives

$$U_0 + \frac{a_3}{a_2 + a_3}\left(H X_0 + \beta_\chi \frac{\partial X_0}{\partial \chi} + \frac{q'}{q} X_0 + (\chi-\chi_0)\frac{q'}{q}\frac{\partial X_0}{\partial \chi} \right) = 0 \ . \tag{A.7}$$

Introducing the abbreviations

$$E = H + \frac{q'}{q}, \quad G = \hat{\beta}_\chi + (\chi-\chi_0)\frac{q'}{q} \ , \tag{A.8}$$

expression (A.6) becomes

$$W_p(n=\infty) = \frac{1}{2}\int_{-\infty}^{+\infty} d\chi \left\{ a_1 \left| \frac{\partial X_0}{\partial \chi} \right|^2 + \frac{a_2 a_3}{a_2 + a_3} \left| E X_0 + G \frac{\partial X_0}{\partial \chi} \right|^2 - a_5 |X_0|^2 \right\} . \tag{A.9}$$

The ballooning-mode criterion consists of checking the sign of W_p $(n=\infty)$ of (A.9) on each flux surface. Equation (A.9) is identical to

$$W_p(n=\infty) = \frac{1}{2}\int_{-\infty}^{+\infty} d\chi \left\{ A \left| \frac{\partial X_0}{\partial \chi} + B X_0 \right|^2 + C |X_0|^2 \right\} , \tag{A.10}$$

or

$$W_p(n=\infty)=\frac{1}{2}\int_{-\infty}^{+\infty} d\chi \left[A\left|\frac{\partial X_0}{\partial\chi}\right|^2 + \left(C+AB^2-\frac{\partial(AB)}{\partial\chi}\right)|X_0|^2\right] , \qquad (A.11)$$

where

$$A=a_1+\frac{a_2a_3}{a_2+a_3}G^2$$

$$A\cdot B=\frac{a_2a_3}{a_2+a_3}EG \qquad (A.12)$$

$$A\cdot C=\frac{a_1a_2a_3}{a_2+a_3}E^2-a_5A ,$$

and

$$C+AB^2-\frac{\partial(AB)}{\partial\chi}=-\frac{4\psi Tr^4}{q(|\nabla\psi|^2+T^2)}p'\left[2\left(\frac{\partial P}{\partial\psi}\right)_n-\frac{Gr^2}{|\nabla\psi|^2+T^2}\frac{\partial P}{\partial\chi}\right] , \qquad (A.13)$$

where

$$P=p+\frac{1}{2r^2}(|\nabla\psi|^2+T^2) . \qquad (A.14)$$

The Euler equations corresponding to (A.11) are those of *Connor* et al. (1978) and *Dobrott* et al. (1977).

In *helical* symmetry the same procedure leads to expression (A.10) with slightly different coefficients indexed with H, which are

$$A_H=a_1+\frac{a_2a_3}{a_2+a_3}G_H^2$$

$$A_HB_H=\frac{a_2a_3}{a_2+a_3}E_HG_H+a_1\frac{\partial\tilde{q}}{\partial\chi}\Big/\tilde{q} \qquad (A.15)$$

$$A_HC_H=\frac{a_1a_2a_3}{a_2+a_3}\left(E_H-G_H\frac{\partial\tilde{q}}{\partial\chi}\Big/\tilde{q}\right)^2-a_5A_H ,$$

where

$$E_H=H+\frac{\partial M}{\partial\chi}+\frac{4s\psi_p h|\boldsymbol{u}|^2}{T}$$

$$G_H=\hat{\beta}_\chi+M \qquad (A.16)$$

$$M=\frac{1}{\tilde{q}}\left[\frac{dq}{ds}(\chi-\chi_0)-\hat{\beta}_z+\alpha\hat{\beta}_\chi\right] .$$

The tokomak expressions are reobtained if $h=\hat{\beta}_z=\alpha=0$ and, consequently, $\tilde{q}=q$.

Numerically we proceed in the following way: We first apply the finite hybrid elements to the variational form of the plasma potential energy (A.10 or 11) for a given flux surface. As a result we obtain a tridiagonal matrix A. This matrix is decomposed in

$$A=LDL^{\mathrm{T}} \ . \tag{A.17}$$

Using Sylvester's theorem we know that, for A regular, the number of negative values in the diagonal matrix D is identical to the number of negative eigenvalues in A. If there is none, the specific flux surface is ballooning stable.

B: Sparse Matrix Techniques to Solve $Ax=\lambda Bx$ in ERATO

B.1 The Problem

The discretization with conforming and non-conforming quadrangular Lagrange and Hermite elements of the variational form of the ideal linear MHD equations, in geometry with one symmetry axis, leads to a huge eigenvalue problem of the form

$$\boldsymbol{A}\boldsymbol{x}=\lambda \boldsymbol{B}\boldsymbol{x} \ , \tag{B.1}$$

where A and B are symmetric and sparse matrices, and B is positive definite, λ is the eigenvalue and $\boldsymbol{x}$ the eigenvector.

In our stability calculations we often calculate only the lowest, i.e., the most unstable, eigenvalue. In addition, we usually know its approximate value. An inverse vector iteration is then a very efficient method to calculate it. Since this method converges towards the eigenvalue lying closest to $\lambda=0$, we have to perform an eigenvalue shift of the form

$$\tilde{\boldsymbol{A}}\boldsymbol{x}=\tilde{\lambda}\boldsymbol{B}\boldsymbol{x} \tag{B.2}$$

where

$$\tilde{A}=A-\lambda_0 B \tag{B.3}$$

and

$$\tilde{\lambda}=\lambda-\lambda_0 \ . \tag{B.4}$$

An inverse vector iteration of the form

$$\begin{aligned} \tilde{\boldsymbol{A}}\boldsymbol{x}^{k+1} &= \boldsymbol{B}\boldsymbol{v}^{k} \\ \boldsymbol{v}^{k+1} &= \boldsymbol{x}^{k+1}/\|\boldsymbol{x}^{k+1}\| \ , \end{aligned} \tag{B.5}$$

where $\boldsymbol{v}^0$ is a normalized initial vector, converges towards the eigenvalue lying closest to $\tilde{\lambda}=0$ or $\lambda=\lambda_0$. The iteration (B.5) is performed until for all j,

$$(|v_j^{k+1}|-|v_j^k|)^2<\varepsilon^2\,\|\boldsymbol{v}^k\|^2\ , \tag{B.6}$$

where $\varepsilon \ll 1$.

B.2 Two Numberings of the Components

Before choosing a method to solve the iteration procedure (B.5) we have to know about the structure of the matrices A and B. In ERATO or HERA we have two-dimensional domains Ω_p (plasma region) and Ω_v (vacuum region) separated by the plasma-vacuum interface Γ_p and bounded by a conducting wall Γ_w (Fig. 5.1). As coordinates we choose a flux surface counter s and a non-orthogonal angle χ such that $s=0$ at the magnetical axis, $s=1$ on Γ_p and $s=2$ on Γ_w.

For the case where Ω_p is subdivided into three s- and four χ-intervals, two different numberings of the nodal points of the displacement components X, V, and Z are shown in Figs. B.1 and B.2, and are denoted by D1 and D2, respectively. Both types of numberings are made for the lowest-order finite hybrid elements and lead to blocked matrices. Two sequential blocks overlap through the subblock due to the vector component x which, in the variational

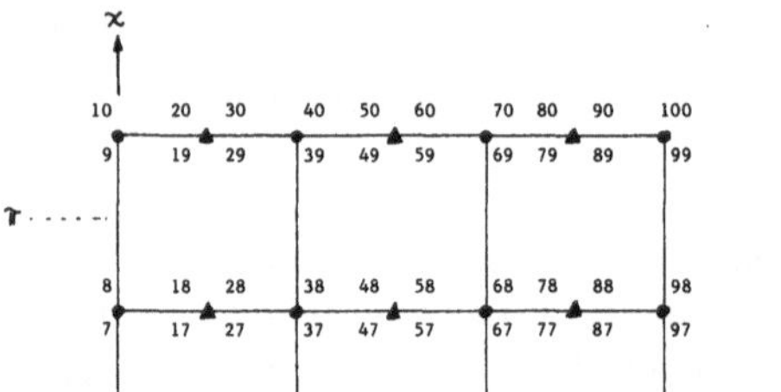

Fig. B.1

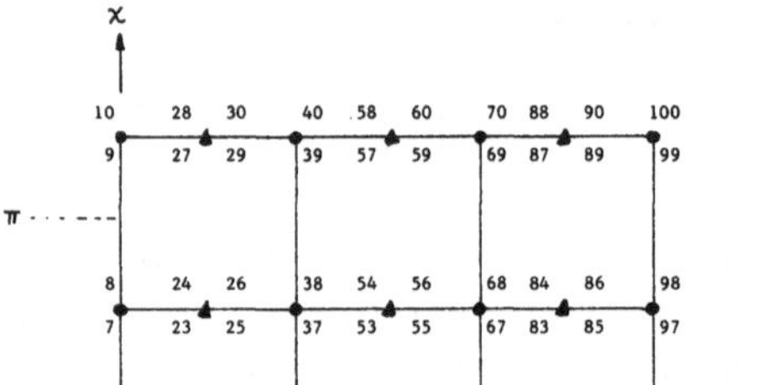

Fig. B.2

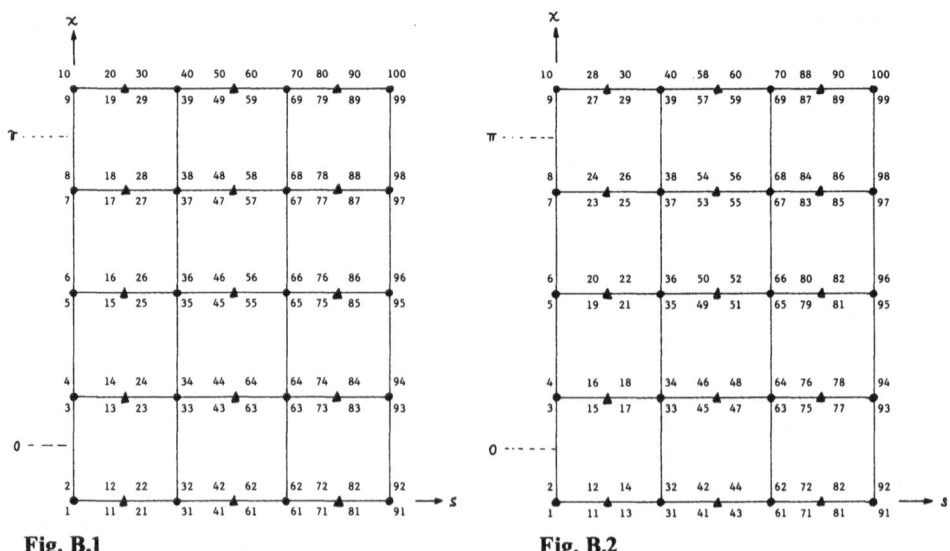

Fig. B.1. Initial numbering D1. Two unknowns per nodal point for real and imaginary part. $N_s=3$, $N_\chi=4$. (●): nodal positions of X, (▲): nodal positions of V and Z

Fig. B.2. Renumbering D2. $N_s=3$, $N_\chi=4$. (●): nodal positions of X, (▲): nodal positions of V and Z

Fig. B.3. Structure of one matrix block for discretization D1

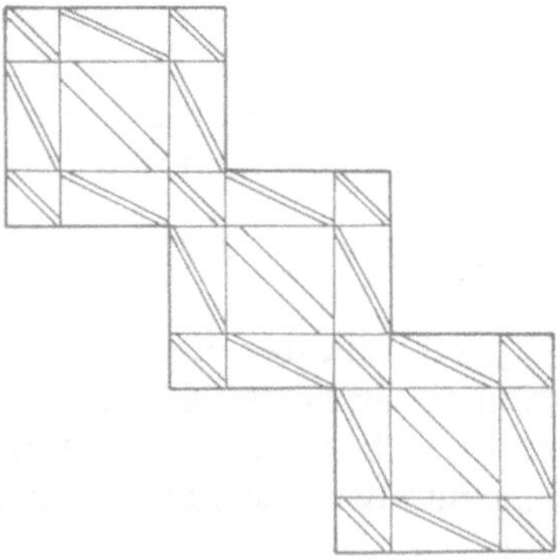

Fig. B.4. Matrix structure for discretization D2

form (5.52–56), is the only one having a derivative in respect to s. With numbering (D1) one block has an internal subblock structure as shown in Fig. B.3. The numbering (D2) leads to a matrix structured as shown in Fig. B.4.

B.3 Resolution for Numbering (D1)

In the case of a matrix structured as shown in Fig. B.3, the eigenvalue problem (B.2) is solved by our eigenvalue solver HYMNIABLOCK (*Gruber* 1980). To solve the system of linear equations (B.5) efficiently, one first performs a matrix decomposition

$$\tilde{A} = LDL^{\mathrm{T}} , \tag{B.7}$$

where L is a regular left-hand side matrix, L^{T} is its transposed, i.e., a right-hand side matrix, and D is a diagonal matrix. The decomposition (B.7) can be done without any pivoting, as long as A and all its dominant submatrices are regular. This is, in general, the case. Until now the decomposition (B.7) always succeeded in all practical applications. Once the decomposition (B.7) is performed, the inverse vector iteration becomes

$$\begin{aligned} LDL^{\mathrm{T}}\boldsymbol{x}^{k+1} &= B\boldsymbol{v}^k = \boldsymbol{u}^k \\ \boldsymbol{v}^{k+1} &= \boldsymbol{x}^{k+1}/\|\boldsymbol{x}^{k+1}\| \ . \end{aligned} \tag{B.8}$$

The matrices $\tilde{A}$ and B are in general very big and sparse. We often calculate with 80 blocks ($N_s = 80$) of size 336×336 ($N_\chi = 41$) each, altogether 20244 unknowns. In the multiplication of B with the vector $\boldsymbol{v}^k$ (B.8) one can make maximal profit of the sparseness of B. However, without any special precautions, the decomposition (B.7) fills in the blocks of L. Each of these blocks have then $(4N_\chi + 4) * (8N_\chi + 9)$ non-zero elements ($= 56616$ for $N_\chi = 41$). There are N_s of such blocks.

When running ERATO together with HYMNIABLOCK on a fast computer such as CRAY 1, the turn-around time is mostly given by the time used to bring from the disks the blocks into memory. The number of words transferred during a resolution of (B.8) is proportional to

$$\mathrm{IO(D1)} \cong N_s(4N_\chi+4)(8N_\chi+9)(3N_{\mathrm{it}}+4) , \tag{B.9}$$

where N_{it} is the number of iterations to be performed. This amount differs from that given in *Gruber* (1980) since we do not further calculate the Rayleigh quotient at each iteration step. For a practical case of $N_s=80$, $N_\chi=41$, and $N_{\mathrm{it}}=5$ the number of words to be transferred from disk to memory is more than 80×10^6!

B.4 Resolution for Numbering (D2)

In the matrix shown in Fig. B.4 the new unknown Y includes the two components V and Z. The system of linear equations for the case with $N_s=3$ intervals is written

$$\begin{aligned}
&A_{11}X_1+A_{12}Y_1+A_{13}X_2 &&=U_1\\
&A_{12}^{\mathrm{T}}X_1+A_{22}Y_1+A_{23}X_2 &&=U_2\\
&A_{13}^{\mathrm{T}}X_1+A_{23}^{\mathrm{T}}Y_1+A_{33}X_2+A_{34}Y_2+A_{35}X_3 &&=U_3\\
&\qquad A_{34}^{\mathrm{T}}X_2+A_{44}Y_2+A_{45}X_3 &&=U_4\\
&\qquad A_{35}^{\mathrm{T}}X_2+A_{45}^{\mathrm{T}}Y_2+A_{55}X_3+A_{56}Y_3+A_{57}X_4 &&=U_5\\
&\qquad\qquad A_{56}^{\mathrm{T}}X_3+A_{66}Y_3+A_{67}X_4 &&=U_6\\
&\qquad\qquad A_{57}^{\mathrm{T}}X_3+A_{67}^{\mathrm{T}}Y_3+A_{77}X_4 &&=U_7 .
\end{aligned} \tag{B.10}$$

In a first step we eliminate the two components V and Z which are compiled in Y:

$$\begin{aligned}
Y_1&=A_{22}^{-1}(U_2-A_{12}^{\mathrm{T}}X_1-A_{23}X_2)\\
Y_2&=A_{44}^{-1}(U_4-A_{34}^{\mathrm{T}}X_2-A_{45}X_3)\\
Y_3&=A_{66}^{-1}(U_6-A_{56}^{\mathrm{T}}X_3-A_{67}X_4) .
\end{aligned} \tag{B.11}$$

Putting expressions (B.11) back into (B.10) leads to

$$\begin{aligned}
&\tilde A_{11}X_1+\tilde A_{13}X_2 &&=\tilde U_1\\
&\tilde A_{13}^{\mathrm{T}}X_1+\tilde A_{33}X_2+\tilde A_{35}X_3 &&=\tilde U_3\\
&\qquad \tilde A_{35}^{\mathrm{T}}X_2+\tilde A_{55}X_3+\tilde A_{57}X_4 &&=\tilde U_5\\
&\qquad\qquad \tilde A_{57}X_3+\tilde A_{77}X_4 &&=\tilde U_7 ,
\end{aligned} \tag{B.12}$$

where

$$\begin{aligned}
\tilde{A}_{11} &= A_{11} - A_{12}A_{22}^{-1}A_{12}^{T} \\
\tilde{A}_{13} &= A_{13} - A_{12}A_{22}^{-1}A_{23} \\
\tilde{A}_{33} &= A_{33} - A_{23}^{T}A_{22}^{-1}A_{23} - A_{34}A_{44}^{-1}A_{34}^{T} \\
\tilde{A}_{35} &= A_{35} - A_{34}A_{44}^{-1}A_{45} \\
\tilde{A}_{55} &= A_{55} - A_{45}^{T}A_{44}^{-1}A_{45} - A_{56}A_{66}^{-1}A_{56}^{T} \\
\tilde{A}_{57} &= A_{57} - A_{56}A_{66}^{-1}A_{67} \\
\tilde{A}_{77} &= A_{77} - A_{67}^{T}A_{66}^{-1}A_{67} \ ,
\end{aligned} \tag{B.13}$$

and

$$\begin{aligned}
\tilde{U}_1 &= U_1 - A_{12}A_{22}^{-1}U_2 \\
\tilde{U}_3 &= U_3 - A_{23}^{T}A_{22}^{-1}U_2 - A_{34}A_{44}^{-1}U_4 \\
\tilde{U}_5 &= U_5 - A_{45}^{T}A_{44}^{-1}U_4 - A_{56}A_{66}^{-1}U_6 \\
\tilde{U}_7 &= U_7 - A_{67}^{T}A_{66}^{-1}U_6 \ .
\end{aligned} \tag{B.14}$$

The only time-consuming operations in (B.13) and (B.14) seem to be the inversions of the matrices A_{22}, A_{44}, and A_{66}. As we can see in Fig. B.4, and magnified for one block in Fig. B.5, all these matrices are banded with a band width of M. Instead of inverting them we have to decompose them into

$$\begin{aligned}
A_{22} &= L_{22}D_{22}L_{22}^{T} \\
A_{44} &= L_{44}D_{44}L_{44}^{T} \\
A_{66} &= L_{66}D_{66}L_{66}^{T} \ .
\end{aligned} \tag{B.15}$$

We then replace inversion by solving for two banded systems. Note that inversion takes a number of operations proportional to $N_2^3/6$ whereas a decomposition (B.15) only implies $\sim M^2N_2/2$ operations. In addition, memory

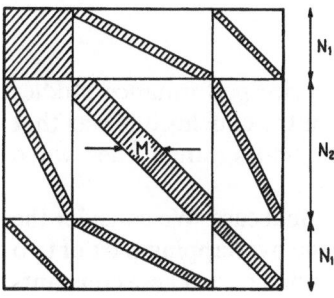

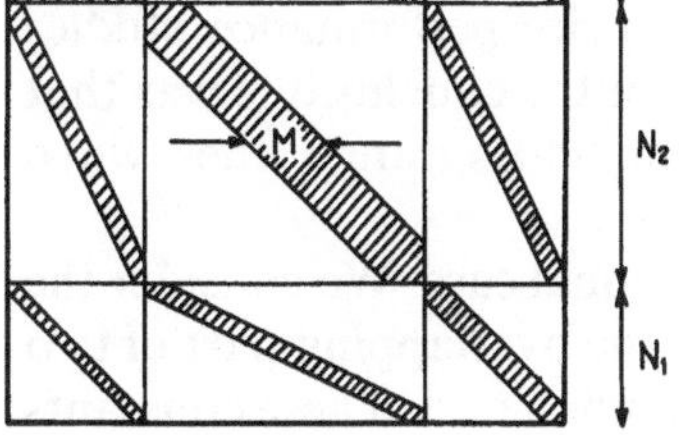

Fig. B.5. One matrix block of discretization D2 during decomposition procedure

space involved in an inversion is proportional to $N_2^2/2$ compared with $MN_2/2$ for a decomposition. For the case with $N_\chi=41$ intervals, and with lowest-order finite hybrid elements $N_2/M=11.2$.

The next step is to solve the system (B.12) to (B.14). This is done by the well-known Gauss elimination procedure

$$\begin{aligned}
&\tilde{\tilde{A}}_{33}=\tilde{A}_{33}-\tilde{A}_{13}^{\mathrm{T}}\tilde{A}_{11}^{-1}A_{13}, \quad \tilde{\tilde{U}}_3=\tilde{U}_3-\tilde{A}_{13}^{\mathrm{T}}\tilde{A}_{11}^{-1}\tilde{U}_1\\
&\tilde{\tilde{A}}_{55}=\tilde{A}_{55}-\tilde{A}_{35}^{\mathrm{T}}\tilde{\tilde{A}}_{33}^{-1}\tilde{A}_{35}, \quad \tilde{\tilde{U}}_5=\tilde{U}_5-\tilde{A}_{35}^{\mathrm{T}}\tilde{\tilde{A}}_{33}^{-1}\tilde{\tilde{U}}_3\\
&\tilde{\tilde{A}}_{77}=\tilde{A}_{77}-\tilde{A}_{57}^{\mathrm{T}}\tilde{\tilde{A}}_{55}^{-1}\tilde{A}_{57}, \quad \tilde{\tilde{U}}_7=\tilde{U}_7-\tilde{A}_{57}^{\mathrm{T}}\tilde{\tilde{A}}_{55}^{-1}\tilde{\tilde{U}}_5 \ .
\end{aligned} \tag{B.16}$$

The only operations which involve decomposition of full matrices are the inversions of $\tilde{A}_{11}$, $\tilde{\tilde{A}}_{33}$ and $\tilde{\tilde{A}}_{55}$.

The back-substitution also involves the inversion of $\tilde{\tilde{A}}_{77}$. It is written

$$\begin{aligned}
&X_4=\tilde{\tilde{A}}_{77}^{-1}\tilde{\tilde{U}}_7\\
&X_3=\tilde{\tilde{A}}_{55}^{-1}(\tilde{\tilde{U}}_5-\tilde{A}_{57}X_4)\\
&X_2=\tilde{\tilde{A}}_{33}^{-1}(\tilde{\tilde{U}}_3-\tilde{A}_{35}X_3)\\
&X_1=\tilde{A}_{11}^{-1}(\tilde{U}_1-\tilde{A}_{13}X_2) \ .
\end{aligned} \tag{B.17}$$

The other two components V and Z of the displacement vector are defined by (B.11).

We have found that such a procedure diminishes the CPU-time by about a factor of 3, on scalar machines. Even though special care has been taken in vectorizing the code, no gain in CPU-time is observed on a vector machine, since case (D1) is better vectorizable than case (D2). The amount of memory space used also diminishes by a factor of 3. Most gain is obtained in the number of words to be transferred from disk to memory. In case (D2)

$$\mathrm{IO(D2)} \cong N_s(N_\chi+1)\,[4N_\chi N_{\mathrm{it}}+318N_{\mathrm{it}}+213] \ , \tag{B.18}$$

giving a reduction by a factor of 8.2 with respect to (D1) for the case $N_s=80$, $N_\chi=41$, and $N_{\mathrm{it}}=5$.

B.5 Higher Order Finite Elements

At the end of Chap. 3 we made comparisons of price-performance studies between different types of 1D-finite elements. One of the conclusions was that Hermite elements were more performant than cubic-Lagrange ones when calculating for a global mode.

The same study is done here for the two-dimensional case. We consider the two numberings (D1) and (D2), and denote by N_1 the overlapping part of two sequential matrix blocks, and by N_2 the non-overlapping part. These constants

Table B.1. Matrix characteristics

Type of element	$N1$	$N2$	M
Lagrange linear L1	$2N_\chi+2$	$4N_\chi+4$	15
Lagrange quadratic L2	$4N_\chi+2$	$20N_\chi+10$	47
Lagrange cubic L3	$6N_\chi+2$	$48N_\chi+16$	95
Hermite cubic H3	$16N_\chi+16$	$8N_\chi+8$	31

are listed in Table B.1 for linear (L1), quadratic (L2) cubic (L3) Lagrange and for Hermite (H3) elements.

For the numbering (D1) we obtain N_s matrix blocks of size $2N_1+N_2$, each with an overlap of N_1 to the previous and the following block. These matrix blocks are considered to be filled in when treating them with HYMNIABLOCK. The CPU-time scales with $N_s*(N_1+N_2)+(2N_1+N_2)^2/2$ and the number of IO operations with $(2N_1+N_2)^2*N_s*N_{it}$. It is evident that for such scaling laws, the Hermite elements (H3) are far more efficient than the cubic Lagrange (L3) ones.

The situation looks different when we compare H3 with L3 for numbering (D2). There the computing time scales with $N_s*(N_1^3+\alpha M^2 N_2)$, where α measures the time of the band matrix operations, relative to the inversion of the full blocks, the memory size with $(N_1+N_2)*N_s+M*N_2+N_1^2$ and the numbering of IO operations with $(M*N_2+N_1^2)*N_s*N_{it}$. In the case of $N_\chi=20$ the two methods (L3 and H3) are about equivalent. The greater the number of N_χ intervals taken, the more L3 overcomes H3.

C: Organization of ERATO

The development costs of a computer code of the size of ERATO (10,000 lines or more) can be compared with those of a major piece of instrumentation. We therefore require that such a program be well structured, and that it consist of interchangeable modules, be easily transportable from one computer to another, efficient, readable, well documented and easy to modify. A big effort has been made to fulfil these requirements in ERATO. To be portable and efficient, ERATO uses STANDARD FORTRAN as the programming language. STANDARD FORTRAN is well defined by ANSI and the computer manufacturers make a big effort to provide the users with very efficient FORTRAN compilers. In addition, on vector machines FORTRAN is the only symbolic language which guarantees an efficient use of the computing power.

Special care has been taken in the structure of ERATO. The equilibrium code and ERATO are distinct codes interlinked through a disk file. The user of ERATO then has the freedom to run his own equilibrium solver. ERATO itself has been subdivided into six main programs which are interlinked through a

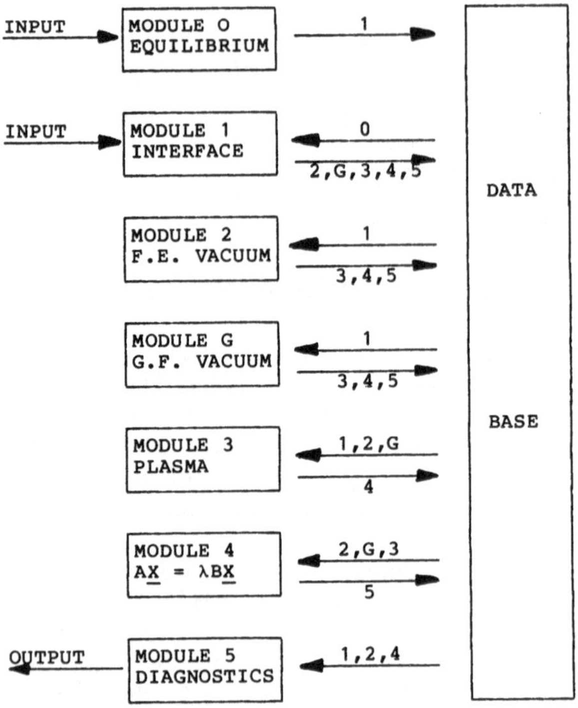

Fig. C.1. Main block structure of ERATO code. Numbers on the arrows denote data transfer from and to other modules

data bank as shown in Fig. C.1. In the first main program ERATO 1 the user has to provide the input through NAMELIST NEWRUN. In this major module the equilibrium input is read from a disk and the information is mapped into a form adequate for the stability calculation. Mercier and ballooning-mode criteria are also checked in this part of the code. This part of ERATO has been organized following the guidelines of the OLYMPUS system (*Roberts* 1974; *Christiansen* and *Roberts* 1974). The second and third main modules, ERATO 2 (or ERATO G) and ERATO 3, produce the discretized vacuum and plasma contributions, respectively. Main block ERATO 4 is the solver of the eigenvalue problem $Ax = \omega^2 Bx$, and ERATO 5 is the diagnostic program. This separation of equilibrium and stability code, and the subdivision of ERATO into six main programs have the advantage that the amount of memory space used is minimized. It also enables researchers who were not involved in the writing of ERATO to replace such main modules on their own. Thus, the mapping part of ERATO 1 has been rewritten by *Takizuka* et al. (1981). *Merkel* (1982) replaced the Green's function vacuum for a toroidal geometry by one for a helically symmetric configuration, and *Scott* (1981) improved the eigenvalue solver ERATO 4 by replacing it by a very efficient sparse matrix solver, described in Appendix B. Each of these main modules has its own modular structure. As illustration, we show in Appendix D, all the COMMONS used in ERATO, and the listing of ERATO 3, which deals with the plasma contribution to the variational form in toroidal geometry, as given in (5.52–57). Each quadratic term in the integrals (5.52 and 55) is treated separately. Any double integral is written

$$\frac{1}{2}\int_0^1\int_0^{2\pi} a_k|I_k|^2 ds\, d\chi = \sum_{i=1}^{N_s}\sum_{j=1}^{N_\chi}\frac{1}{2}\iint_{\text{cell}} w_{ij}a_k|I_k|^2 ds\, d\chi, \tag{C.1}$$

where

$$w_{ij}=\begin{cases}1 & \text{if} \quad 1<j<N_\chi \\ 1/2 & \text{if} \quad j=1 \quad \text{or} \quad j=N_\chi \;.\end{cases} \tag{C.2}$$

The summation over i (=JPSI in the code) and j (=JCHI) is made in subroutine CONMAT, which calls subroutine INTEGR. Due to the use of lowest-order finite hybrid elements for the discretization of the displacement components X, V, and Z, the integration over one mesh cell is identical to multiplying the value of the integrand at the center of the cell with the mesh area, i.e.,

$$\iint_{\text{cell}} w_{ij}a_k|I_k|^2\, ds\, d\chi = w_{ij}a_k|I_k|^2(s_{i-1/2},\chi_{j-1/2})(s_i-s_{i-1})(\chi_j-\chi_{j-1}) \;. \tag{C.3}$$

The integrand $a_k|I_k|^2$ consists of 16×16 matrix contributions (Figs. B.1 and B.2) since there are 16 unknowns giving non-zero matrix contributions in a cell. In addition, there are altogether 16 different basis functions for the function values and their derivatives (four basis functions for the function, the s-derivative and the χ-derivatives of X, and two basis functions for the function and the χ-derivatives of V or Z). The 16×16 matrices for the potential and kinetic energies are constructed as follows: In the two subroutines AHYBRD and BHYBRD called by INTEGR the cell contributions to the plasma potential and kinetic energies are calculated. After having initialized to zero the cell contribution matrix, the 16 basis function values at the center of the cell are computed in subroutine BASIS 2. In DELW 1 the coefficients a_k of the quadratic terms are calculated. In DELW 2 and DELW 3 the different quadratic terms with and without a variable transformation are calculated, as presented in Sect. 5.6. In VECT 2 the vector of the coefficients of the 16 unknowns is constructed. To obtain the final 16×16 matrix, a diadic multiplication of this vector with its complex conjugate, as well as a scaling with the corresponding coefficient has to be performed. These operations, as well as the addition of the quadratic terms, are done in subroutine DIADC 2. The total contribution of a mesh cell is obtained when multiplication with the surface area is accomplished.

Once the mesh-cell contribution is calculated, we have to introduce symmetry conditions by calling SYMTRY for $j=1$ and $j=N_\chi$, regularity conditions by calling REGULA for $i=1$ and boundary conditions by calling BOUNDA for $i=N_s$. Finally the mesh contribution including all the conditions is added to the big matrix in subroutine STORE.

Proceeding in such a way enables us to rearrange easily the plasma potential energy by simply rewriting the subroutines DELW 1, DELW 2, and DELW 3. If we have to change the symmetry, regularity or boundary conditions we only have to touch the specific subroutines. We must also mention here that all input-output operations are collected in one subroutine (IODSK 3), and that the numbering of the unknowns is defined in DEC. This makes it possible to renumber the unknowns easily, as shown in Appendix B.

Publishing numerical results is similar to publishing experimental results. The reader of such a publication relies on the integrity of the authors, since it is often not possible to check the given results. Errors can be made in the construction of an experiment, in the diagnostics and in the interpretation of the obtained raw data. The same can happen in computer codes. Errors in programming, in the choice of the numerical method and in the interpretation of the obtained results can be made.

In ERATO we took special care to obtain a well-tested robust software giving easily interpretable results. A test has been made by comparing its results with 1D results produced by THALIA (*Appert* et al. 1975b), and with 2D fixed boundary results obtained by Kerner and the PEST code (*Chance* et al. 1978). A major problem was to make sure that the Green's function vacuum part described in *Gruber* et al. (1981a) was correct. For this purpose we have entirely rewritten the vacuum part and we solve it again by a finite element approach very similar to that for the plasma region. In this method, one considers the vacuum as a pressureless and currentless pseudoplasma. We found that both methods for the discretization of the vacuum contribution give the same results (*Gruber* et al. 1981b). Due to this cross check, we were able to recheck most of the plasma contribution terms.

If a computer code is supposed to be made public, it is necessary to provide a complete documentation. Thus, an earlier version of ERATO was published in the Belfast program library (*Gruber* et al. 1981a). The extensions made since then have been reported in *Gruber* et al. (1981b).

D: Listing of ERATO 3 (with R. Iacono)

```
*AF ERA3NEW,P1C3S09
*COMDECK COMBAS
C-------------------------------------------------------------------------
CL                        C1.1.      BASIC SYSTEM PARAMETERS
C      VERSION 2C               1.8.73        KVR/MHH           CULHAM
        COMMON/COMBAS/
C-------------------------------------------------------------------------
     R    ALTIME,    CPTIME,    STIME,
     I    NDIARY,    NIN,       NLEDGE,    NONLIN,    NOUT,      NPRINT,
     I    NPUNCH,    NREAD,     NREC,      NRESUM,    NRUN,      NSTEP,
     L    NLEND,     NLRES,
     H    LABEL1,    LABEL2,    LABEL3,    LABEL4,    LABEL5,    LABEL6,
     H    LABEL7,    LABEL8
        LOGICAL
     L    NLEND,     NLRES
        DIMENSION
     H    LABEL1(5),            LABEL2(5),            LABEL3(5),
     H    LABEL4(5),            LABEL5(5),            LABEL6(5),
     H    LABEL7(5),            LABEL8(5)
```

```
*COMDECK COMPHY
C-----------------------------------------------------------------------
CL                C2.1        GENERAL PHYSICS VARIABLES
C       VERSION 2C            14/9/79     RG       CRPP LAUSANNE
      COMMON/COMPHY/
C-----------------------------------------------------------------------
     R    ALO,        ASPCT,      BETA,       BETAP,      BETAS,      B2R2,
     R    CDQ,        CIQ,        CMERC,      CPPR,       CPR,        CPSRF,
     R    CST,        DQ,         ELLIPT,     GAMMA,      Q,          QIAXE,
     R    QPSI,       QS,         QSURF,      QTILDA,     REXT,       RHO,
     R    RSUR,       SB2,        SCALE,      SI,         SP,         SP2,
     R    SV,         TMF,        TQP,        TSURF,      TTP,        WNL,
     R    WNTORE
      DIMENSION
     R    CDQ(    15),            CIQ(    15),            CMERC(    15),
     R    CPPR(    15),           CPR(    15),            DQ(    3),
     R    Q(    3),               QS(    61),             QPSI(    15)
     R    QTILDA(    15),         RHO(    15),            TMF(    15),
     R    TQP(    15),            TTP(    15)
*COMDECK COMEQU
C-----------------------------------------------------------------------
CL                C2.2        EQUILIBRIUM VARIABLES AND ARRAYS
C       VERSION 2C            14/9/79     RG       CRPP LAUSANNE
      COMMON/COMEQU/
C-----------------------------------------------------------------------
     R    EQ,
     I    IEQ
      DIMENSION
     R    EQ( 20,  15)
*COMDECK COMEIG
C-----------------------------------------------------------------------
CL                C2.3        EIGENVALUE PARAMETERS
C     VERSION 1C           12/4/75     DB/RG           CRPP LAUSANNE
      COMMON/COMEIG/
C-----------------------------------------------------------------------
     R    AL,         EPSCON,     EPSMAC,
     I    NITMAX,     NPIN,       NPOUT
      DIMENSION
     R    AL(    3),
     I    NPIN(   14),            NPOUT(    4)
*COMDECK COMVEC
C
C-----------------------------------------------------------------------
CL                C2.4        VECTORS FOR INVERSE ITERATION
C
          COMMON/COMVEC/
C-----------------------------------------------------------------------
     R    X,          U,          V
      DIMENSION
     R    X( 2684),   U( 264),    V( 264)
*COMDECK COMESH
C-----------------------------------------------------------------------
CL                C3.1        MESH VARIABLES
C       VERSION 2C            14/9/79     RG       CRPP LAUSANNE
      COMMON/COMESH/
C-----------------------------------------------------------------------
     R    APLACE,     AWIDTH,     CHI,        CPSI,       CR,         CS,
     R    CZ,         DPDR,       DPDZ,       RPLT,       SMAX,       SOLPDS,
     R    WIDTH,
     I    NCHI,       NPOIDS,     NPSI,       NSHIFT,     NV
      DIMENSION
     R    APLACE(    10),         AWIDTH(    10),         CHI (    16),
     R    CPSI(    15),           CR(    15),             CS(    15),
     R    CZ(    15),             DPDR(    15),           DPDZ(    15),
     R    RPLT(   402)
*COMDECK COMNUM
C-----------------------------------------------------------------------
CL                C3.2        NUMERICAL VARIABLES
C       VERSION 2C            14/9/79     RG       CRPP LAUSANNE
      COMMON/COMNUM/
C-----------------------------------------------------------------------
     R    AI1,        AI2,        AJR,        AR,         AZ,         BP,
     R    CPI,        DBP2,       DPR,        DPZ,        FIT,        PSILIM,
     R    XINT1,      XINT2,
     I    LENGTH,     NAN,        NANL,       NANR,       NANZ,       NCOLMN,
     I    NR,         NTURN,      NUM,        NZ
      DIMENSION
     R    AR(   201),             AZ(   201),             FIT(    16),
     R    XINT1(   201),          XINT2(   201)
*COMDECK COMIVI
C-----------------------------------------------------------------------
CL                C3.3        INVERSE VECTOR ITERATION VARIABLES
C       VERSION 2C            14/9/79     RG       CRPP LAUSANNE
      COMMON/COMIVI/
C-----------------------------------------------------------------------
     R    ALO,        ALAM,       XNORM,
     I    M1,         M11,        M2,         M12,        N,          NCONV,
     I    NEG,        NIT,        NITMAX,     NLONG,      NSING,      NTYPE
```

```
*COMDECK COMCON
C-----------------------------------------------------------------------
CL                    C4.1         CONTROL VARIABLES
C         VERSION 2C               14/9/79        RG        CRPP LAUSANNE
      COMMON/COMCON/
C-----------------------------------------------------------------------
     I   NMESH,     NTCASE,
     L   NLEING,    NLGREN,     NLPLOT,    NLTORE,    NLY
      LOGICAL
     L   NLEING,    NLGREN,     NLPLOT,    NLTORE,    NLY
*COMDECK COMOUT
C-----------------------------------------------------------------------
CL                    C5.1         OUTPUT CHANNEL NUMBERS
C         VERSION 2C               14/9/79        RG        CRPP LAUSANNE
      COMMON/COMOUT/
C-----------------------------------------------------------------------
     R   ALARG,     ANGLE,      ARROW,     CMAX,      PMAX,      PPMAX,
     R   GMAX,      TTPMAX,
     I   ITEST,     MEG,        NDA,       NDB,       NDES,      NDLT,
     I   NDORY,     NDS,        NDSCRC,    NFIG,      NPRNT,     NSAVE,
     I   NSCRTC,    NSUR,       NVAC
      DIMENSION
     R   ANGLE(   16)
*COMDECK COMMAP
C-----------------------------------------------------------------------
CL                    C9.1         BLANK COMMON FOR ERATO1
C
      COMMON
C-----------------------------------------------------------------------
     R   A0,        A1,         A2,        A3,        A4,        A5,
     R   A6,        A7,         A8,        A9,        APP,       ATTP,
     R   BETRZ,     CHIRZ,      CPSEQ,     EQPS,      EQR,       EQT,
     R   EQZ,       GPSIP,      RAXIS,     RMAG,      RPSI1,     RRO,
     R   RRS,       RZS,        TETA,      TLIM,      TO,
     I   ILOC,      NRZ,        NRZS
      DIMENSION
     R   A0(  131),             A1(  131),            A2(  131),
     R   A3(  131),             A4(  131),            A5(  131),
     R   A6(  131),             A7(  131),            A8(  131),
     R   A9(  131),             ABAL(    1),          APP(10),
     R   ATTP(    10),          BETRZ(  131),         CHIRZ(  131),
     R   CPSEQ(  65,   33),     EQPS(  65),           EQR(    65),
     R   EQT(    65),           EQZ(    33),          GPSIP(  131),
     R   RRO(  131),            RRS(  131),           RZS(   131),
     R   TETA(  131),
     I   ILOC(    15),          NBAL(    1)
      EQUIVALENCE
     R     (ABAL,CPSEQ),         (NBAL,EQR)
*COMDECK COMVAC
C-----------------------------------------------------------------------
CL                    C9.2                   BLANK COMMON FOR ERATO2
C
      COMMON
C-----------------------------------------------------------------------
     R   APLACE,    AWIDTH,     BPS,       CHI,       CPI,       CPSRF,
     R   CR,        CS,         CZ,        DPDR,      DPDZ,      DRDTCW,
     R   DRDTPS,    QIAXE,      GPSI,      QS,        QSURF,     QTILDA,
     R   RO,        REXT,       ROCW,      ROPS,      SMAX,      TS,
     R   TSURF,     TV,         WNL,
     I   NCHI,      NPSI,       NV
      DIMENSION
     R   APLACE(  10),          AWIDTH(   10),        BPS(   10),
     R   CHI( 102),             CR( 102),             CS( 102),
     R   CZ( 102),              DPDR( 102),           DPDZ( 102),
     R   DRDTCW(  102),         DRDTPS(  102),        GPSI( 150),
     R   QS( 300),              QTILDA( 150),         ROCW(102),
     R   ROPS(102),             TV( 102)
```

```
*COMDECK COMVID
C
C--------------------------------------------------------------------------
CL                 C9.G        VACUUM QUANTITIES FOR ERATOG
C
      COMMON
C--------------------------------------------------------------------------
     R  CA,        CB,         CC,         DC,         DRODT,      DROPDT,
     R  D2ROP,     G,          GB,         RO,         ROP,        RO2,
     R  RO2P,      SR,         SZ,         T,          TB,         TH,
     R  TP,        TT,
     I  NDIM
      DIMENSION
     R  A(   28,   28),        B(   28,   28),         C(   28,   28),
     R  D(   28,   28),        DC(   32),              DRODT(   4),
     R  DROPDT(   4),          E(   28,   28),         F(   28,   28),
     R  G(  28,   28,  6),     H(  784),               RO(   4),
     R  ROP(   4),             SR(   32),              SZ(   32),
     R  T(   32),              TH(   32),              TP(   4),
     R  TT(   4),              WVAC(   58,   58)
        EQUIVALENCE
     1  (A(1,1),G(1,1,1)),     (B(1,1),G(1,1,2)),      (C(1,1),G(1,1,3)),
     2  (D(1,1),G(1,1,4)),     (E(1,1),G(1,1,5)),      (F(1,1),G(1,1,6)),
     3  (WVAC(1,1),G(1,1,2)),                          (H(1),A(1,1))
*COMDECK COMAUX
C--------------------------------------------------------------------------
CL                 C9.3        BLANK COMMON FOR ERATO3
C       VERSION 2C             14/9/79       RG        CRPP LAUSANNE
      COMMON
C--------------------------------------------------------------------------
     R  AA,        BB,         CONA,       CONB,       XC,         XC1,
     R  XF,        XV,
     I  NPLAC
      DIMENSION
     R  AA( 18,132),           BB( 18,132),            CONA( 16, 16),
     R  CONB( 16, 16),         VAC( 990),              XC(14,16),
     R  XC1(   16),            XF(   16),              XV(16,16),
     I  NPLAC(   16)
*COMDECK COMMTR
C--------------------------------------------------------------------------
CL                 C9.4        STORAGE OF MATRIX BLOCKS FOR ERATO4
C       VERSION 2C             14/9/79       RG        CRPP LAUSANNE
      COMMON
C--------------------------------------------------------------------------
     R  AB,        ALBAND,     CA,         CA2,        OFDIAG
      DIMENSION
     R  AB( 18, 132),          ALBAND( 8, 88),         CA( 990),
     R  CA2( 990),             OFDIAG( 5, 7, 44),      XT(   1)
      EQUIVALENCE
     R  (CA2,XT)
*COMDECK COMPLO
C --------------------------------------------------------------------------
CL                 C9.5        OUTPUT AND PLOT VARIABLES FOR ERATO5
C       VERSION 2C             14/9/79       RG        CRPP LAUSANNE
        COMMON //
C--------------------------------------------------------------------------
     R  CHI,       CNR,        CNZ,        CR,         CZ,         ECHEL,
     R  PHIP,      RCHI,       RCHII,      RCHIR,      RCHR,       RPHII,
     R  RPHIR,     RPSI,       RPSR,       RXI,        RXR,        RYI,
     R  RYR,       RZI,        RZR,        X,          XS,         YS,
     I  KK,        KW,         LABL1,      LABL2,      NCHI,       NDX,
     I  NPSI,      NSHIFT,     NV
      DIMENSION
     R  APLACE(   10),         AWIDTH(   10),          CHI(   24),
     R  CNR(   15,   16),      CNZ(   15,   16),       CR(   15,   16),
     R  CZ(   15,   16),       DW(  14,  16,  6),      RCHI(   15,   16),
     R  RCHII(   15,   16),    RCHIR(   15,   16),     RCHR(   15,   16),
     R  RK(   15,   16),       RPHII(   15,   16),     RPHIR(   15,   16),
     R  RPSI(   15,   16),     RPSR(   15,   16),      RW(   15,   16),
     R  RXI(   15,   16),      RXR(   15,   16),       RYI(   15,   16),
     R  RYR(   15,   16),      RZI(   15,   16),       RZR(   15,   16),
     R  SUM(  15,   6),        X( 1376),               XS(  262),
     R  YS(  262),
     I  KK(   15,   16),       KW(   15,   16),
     I  KXI(   15,   16),      KXR(   15,   16),       KYI(   15,   16),
     I  KYR(   15,   16),      KZI(   15,   16),       KZR(   15,   16),
     I  LABL1(   5),           LABL2(   5)
      EQUIVALENCE
     R  (RK,KK),               (RW,KW),                (DW,X),
     I  (SUM,CR),
     I  (KXI,RXI),             (KXR,RXR),              (KYI,RYI),
     I  (KYR,RYR),             (KZI,RZI),              (KZR,RZR)
```

```
*COMDECK NEWRUN
C-----------------------------------------------------------------------
CL                     N           NAMELIST
C      VERSION 1C               12/4/75       DB/RG            CRPP LAUSANNE
        NAMELIST /NEWRUN/
C-----------------------------------------------------------------------
      R    ALARG,     ALO,        ANGLE,      APLACE,     ARROW,      ASPCT,
      R    AWIDTH,    B2R2,       CMAX,       CPSRF,      CST,        ELLIPT,
      R    EPSCON,    EPSMAC,     GAMMA,      PMAX,       PPMAX,      QIAXE,
      R    QMAX,      QPSI,       QS,         QSURF,      QTILDA,     REXT,
      R    SCALE,     SMAX,       SOLPDS,     TSURF,      TTPMAX,     WIDTH,
      R    WNL,       WNTORE,
      I    ITEST,     LENGTH,     MEQ,        NAN,        NCHI,       NCOLMN,
      I    NDA,       NDB,        NDLT,       NDS,        NFIG,       NIN,
      I    NITMAX,    NMESH,      NPRNT,      NPOIDS,     NPSI,       NR,
      I    NSAVE,     NSCRTC,     NSHIFT,     NSUR,       NTCASE,     NTURN,
      I    NV,        NVAC,       NZ,
      L    NLEING,    NLGREN,     NLPLOT,     NLTORE,     NLY
C-----------------------------------------------------------------------
```

```
*AF ERA3NEW,P2C3S01
*DECK P3C0S01
      PROGRAM ERATO3(INPUT=70,OUTPUT=70,OUT=70,TA=514,TB=514,
     +   TAPE4=514,TAPE5=INPUT,TAPE6=OUTPUT,TAPE7=TA,TAPE8=OUT,
     +   TAPE9=TB,TAPE10=514,TAPE16=514,TAPE17=514)
C
C        ERATO3 FILLS IN THE PLASMA ENERGY PART OF THE
C        MATRICES A AND B
C
      CALL AANDB
C
      STOP
      END
*DECK P3C2S01
      SUBROUTINE AANDB
C
C 3.2.1    ORGANIZE MATRIX CONSTRUCTION
C
C
C      READ NAMELIST
C
      CALL IODSK3(1,1)
C
C      CREATE BLOCKS OF MATRICES A AND B
C
      CALL CONMAT
      CALL IODSK3(5,1)
C
      STOP
      END
*DECK P3C2S02
      SUBROUTINE CONMAT
C
C 3.2.2    CREATE BLOKS OF A AND B
C
*CALL COMEQU
*CALL COMAUX
*CALL COMESH
*CALL COMOUT
C
      NPSI=NPSI+NV
      CALL RESETR(AA,126*(2*NCHI+2),0.0)
      CALL RESETR(BB,126*(2*NCHI+2),0.0)
      REWIND MEQ
      DO 300 JPSI=1,NPSI
C
C    READ EQUILIBRIUM QUANTITIES
C
      CALL IODSK3(6,JPSI)
C
C    FILL ONE BLOCK
C
      DO 100 JCHI=1,NCHI
      CALL INTEGR(JPSI,JCHI)
 100  CONTINUE
C
C    WRITE BLOCK ON DISK
C
      IF (JPSI .NE. 1) CALL IODSK3(4,JPSI)
      J8=8*NCHI+8
C
C    STORE OVERLAPPING PART
C
      DO 200 J=1,J8
      DO 200 I=1,18
      J6=J+6*NCHI+6
      AA(I,J)=AA(I,J6)
      BB(I,J)=BB(I,J6)
      AA(I,J6)=0.0
      BB(I,J6)=0.0
 200  CONTINUE
 300  CONTINUE
C
C    WRITE LAST BLOCK ON DISK
C
      CALL IODSK3(4,NPSI+1)
C
      RETURN
      END
```

```
*DECK P3C2S03
      SUBROUTINE INTEGR(KPSI,KCHI)
C
C 3.2.3   INTEGRATE OVER ONE CELL
C
*CALL COMPHY
*CALL COMEQU
*CALL COMAUX
*CALL COMESH
*CALL COMNUM
*CALL COMCON
*CALL COMOUT
C
C     DETECT NUMBERING SEQUENCE FOR CELL KCHI
C
      CALL DEC(KCHI,NCHI,NPLAC)
C
C     HYBRID ELEMENTS
C
      CALL MATRIX(KCHI)
C
C     PERFORM EIGENVALUE SHIFT
C
      DO  100  IC=1,16
      DO  100  JC=1,16
      CONA(IC,JC)=CONA(IC,JC)-ALO*CONB(IC,JC)
 100  CONTINUE
C
C     SYMMETRY CONDITIONS
C
      CALL SYMTRY(1,NCHI,KCHI,KPSI)
      CALL SYMTRY(2,NCHI,KCHI,KPSI)
C
C     REGULARITY CONDITIONS
C
      IF ( KPSI.NE.1 ) GO TO 200
      CALL REGULA(1)
      CALL REGULA(2)
 200  IF(REXT.GT.1.0.AND.NLGREN) GO TO 300
C
C     BOUNDARY CONDITIONS
C
      IF ( KPSI.NE.NPSI ) GO TO 300
      CALL BOUNDA(1)
      CALL BOUNDA(2)
 300  CONTINUE
C
C     ADD TO MATRIX BLOCK
C
      CALL STORE(CONA,CONB)
C
      RETURN
      END
*DECK P3C2S04
      SUBROUTINE DEC(K,KM,KPLAC)
C
C 3.2.4 SEARCH FOR MATRIX ADDRESSES ASSOC. WITH A CELL
C
      DIMENSION KPLAC(16)
*CALL COMCON
C
      ID=2*KM+2
      I1=2*K-2
      I2=I1+ID
      I3=I2+ID
      IF (.NOT. NLY) I3=I2
      I4=I3+ID
      IV=4
      DO  100  I=1,IV
      KPLAC(I)   =I1+I
      KPLAC(I+4) =I4+I
      KPLAC(I+8) =I2+I
      IF (.NOT. NLY) KPLAC(I+8)=0
      KPLAC(I+12)=I3+I
 100  CONTINUE
C
      RETURN
      END
```

```
*DECK P3C2S05
      SUBROUTINE MATRIX(KCHI)
C
C 3.2.5   FILL IN MATRICES CONA AND CONB FOR ONE CELL
C
*CALL COMAUX
*CALL COMCON
*CALL COMEQU
*CALL COMESH
C
      CALL RESETR(CONA,16*16,0.0)
      CALL RESETR(CONB,16*16,0.0)
C
C     BASIS FUNCTION VALUES AT POINT KINT
C
      ZX=EQ(5,KCHI)
      ZY=EQ(6,KCHI)
      ZDX=EQ(3,KCHI)-EQ(1,KCHI)
      ZDY=EQ(4,KCHI)-EQ(2,KCHI)
      CALL BASIS2(ZX,ZDX,ZY,ZDY,XF)
C
C     THE CONSTANTS VECTOR FOR ALL TERMS
C
      CALL DELW1(KCHI,XC1)
C
C     THE 14 CONSTANTS FOR ALL 16 QUADRATIC TERMS
C
      IF(.NOT NLEINQ) CALL DELW2(KCHI,XC)
      IF(NLEINQ) CALL DELW3(KCHI,XC)
C
C     FORM VECTOR XV
C
      CALL VECT2(XC,XF,XV)
C
C     DIADIC MULTIPLICATION
C
      CALL DIADC2(16,XV,XC1)
C
C     MULTIPLY WITH AEREA OF INTERVAL
C
      ZW=1.0
      IF (KCHI .EQ. 1 .OR. KCHI .EQ. NCHI) ZW=0.5
      DO 200 I=1,16
      DO 200 J=1,16
      CONA(I,J)=ZW*ZDX*ZDY*CONA(I,J)
      CONB(I,J)=ZW*ZDX*ZDY*CONB(I,J)
  200 CONTINUE
C
      RETURN
      END
```

```
*DECK P3C2S06
      SUBROUTINE BASIS2(PX,ZDX,PY,ZDY,PF)
C
C 3.2.6   HYBRID ELEMNTS AND THEIR DERIVATIVES
C
      DIMENSION PF(16)
C
C     DEFINITION OF BASIS FUNCTIONS PF
C
C     POINT DXDC     X    DXDS  DVDC    V
C       1     1      2     3
C       2     4      5     6
C       3                         7     8
C       4                         9    10
C       5    11     12    13
C       6    14     15    16
C
C     DEFINITION OF POSITION OF THE POINTS
C
C         2---4---6
C         /   /   /
C         /   /   /
C         /   /   /
C         1---3---5
C
C     ALL FUNCTION VALUES
C
      PF(2)=0.25
      PF(5)=0.25
      PF(8)=0.5
      PF(10)=0.5
      PF(12)=0.25
      PF(15)=0.25
C
C     ALL CHI-DERIVATIVES
C
      PF(1)=-0.5/ZDY
      PF(4)=-PF(1)
      PF(7)=-1.0/ZDY
      PF(9)=-PF(7)
      PF(11)=PF(1)
      PF(14)=PF(4)
C
C     ALL S-DERIVATIVES
C
      PF(3)=-0.5/ZDX
      PF(6)=PF(3)
      PF(13)=-PF(3)
      PF(16)=PF(13)
C
      RETURN
      END
```

```
*DECK P3C2S07
      SUBROUTINE DELW1(KCHI,PC1)
C
C 3.2.7   CONSTANTS A TO H OF MULTIPLYING QUADRATIC TERMS
C
*CALL COMEQU
*CALL COMPHY
C
      DIMENSION PC1(16)
C
      ZS=EQ(5,KCHI)
      ZRHO=EQ(7,KCHI)
      ZP=EQ(8,KCHI)
      IF (ZP .EQ. 0.0) ZP=1.E-5
      ZBT=EQ(9,KCHI)
      ZBP=EQ(12,KCHI)
      ZR2=EQ(14,KCHI)
      ZK=EQ(18,KCHI)
      ZJAC=EQ(19,KCHI)
C
C     POTENTIAL ENERGY CONSTANTS ,EQS. 8.3
C
      PC1(1)=ZJAC**3*ZBP/(QIAXE*ZS*ZR2**2)
      PC1(2)=PC1(1)
      PC1(3)=ZJAC*ZBT**2/(4.0*ZS*CPSRF)
      PC1(4)=PC1(3)
      PC1(5)=ZJAC*QIAXE*ZS*ZR2/(4.0*ZBP)
      PC1(6)=PC1(5)
      PC1(7)=ZJAC*ZP*ZR2/(4.0*ZS*QIAXE)
      PC1(8)=PC1(7)
      PC1(9)=-ZK*ZJAC*ZR2/(QIAXE*ZS)
      PC1(10)=PC1(9)
      IF(EQ(8,KCHI) . GT. 0.0) GO TO 100
C
C     MODIFIED CONSTANTS FOR THE VACUUM REGION
C
      PC1(1)=PC1(1)*(TSURF/ZBT)**2
      PC1(2)=PC1(1)
      PC1(3)=PC1(3)*(TSURF/ZBT)**2
      PC1(4)=PC1(3)
      PC1(5)=PC1(5)*(TSURF/ZBT)**2
      PC1(6)=PC1(5)
  100 CONTINUE
C
C     KINETIC ENERGY CONSTANTS , EQS. 8.4
C
      PC1(11)=ZRHO*ZBP*ZJAC/(ZS*QIAXE)
      PC1(12)=PC1(11)
      PC1(13)=ZRHO*ZS*ZR2**3*QIAXE/(4.0*ZJAC*ZBP)
      PC1(14)=PC1(13)
      PC1(15)=ZRHO*ZBT**2*ZR2**2/(4.0*ZS*CPSRF*ZJAC)
      PC1(16)=PC1(15)
C
      RETURN
      END
```

```
*DECK P3C2S08
      SUBROUTINE DELW2(KCHI,PC)
C
C 3.2.8   QUADRATIC TERMS FOR NLEING=.F.
C
*CALL COMEQU
*CALL COMPHY
C
      DIMENSION PC(14,16)
C
C     THE CONSTANTS ARE ARRANGED IN THE FOLLOWING WAY
C
C     DXDCR,DXDCI,XR,XI,DXDSR,DXDSI,DVDCR,DVDCI,VR,VI,DYDCR,DYDCI,YR,YI
C        1     2   3  4    5     6     7     8    9 10    11    12 13 14
C
      CALL RESETR(PC,14*16,0.0)
C
C   EQUILIBRIUM QUANTITIES
C
      ZNG=WNTORE*EQ(11,KCHI)/QIAXE
      ZBETC=EQ(13,KCHI)
      ZH=EQ(17,KCHI)
      ZDRS=EQ(15,KCHI)
      ZDRC=EQ(16,KCHI)
C
C     FIRST  SQUARE OF POTENTIAL ENERGY
C
      PC(1,1)=1.0
      PC(4,1)=ZNG
      PC(2,2)=1.0
      PC(3,2)=-ZNG
C
C     SECOND SQUARE OF POTENTIAL ENERGY
C
      PC(5,3)=1.0
      PC(7,3)=1.0
      PC(6,4)=1.0
      PC(8,4)=1.0
C
C     THIRD  SQUARE OF POTENTIAL ENERGY
C
      PC(1,5)=ZBETC
      PC(3,5)=ZH
      PC(4,5)=ZNG*ZBETC
      PC(5,5)=1.0
      PC(10,5)=-ZNG
      PC(2,6)=ZBETC
      PC(3,6)=-ZNG*ZBETC
      PC(4,6)=ZH
      PC(6,6)=1.0
      PC(9,6)=ZNG
C
C     FOURTH SQUARE OF POTENTIAL ENERGY
C
      PC(3,7)=ZDRS
      PC(5,7)=1.0
      PC(7,7)=1.0
      PC(9,7)=ZDRC
      PC(11,7)=1.0
      PC(13,7)=ZDRC
      PC(14,7)=ZNG
      PC(4,8)=ZDRS
      PC(6,8)=1.0
      PC(8,8)=1.0
      PC(10,8)=ZDRC
      PC(12,8)=1.0
      PC(13,8)=-ZNG
      PC(14,8)=ZDRC
C
C     FIFTH  SQUARE OF POTENTIAL ENERGY
C
      PC(3,9)=1.0
      PC(4,10)=1.0
C
C     FIRST SQUARE OF KINETIC ENERGY
C
      PC(3,11)=1.0
      PC(4,12)=1.0
C
C     SECOND SQUARE OF KINETIC ENERGY
C
      PC(3,13)=-ZBETC
      PC(9,13)=1.0
      PC(13,13)=1.0
      PC(4,14)=-ZBETC
      PC(10,14)=1.0
      PC(14,14)=1.0
C
C     THIRD SQUARE OF KINETIC ENERGY
C
      PC(13,15)=1.0
      PC(14,16)=1.0
      RETURN
      END
```

```
*DECK P3C2S09
      SUBROUTINE DELW3(KCHI,PC)
C
C 3.2.9   QUADRATIC TERMS FOR NLEING=.T.
C
*CALL COMEQU
*CALL COMPHY
C
      DIMENSION PC(14,16)
C
C     THE CONSTANTS ARE ARRANGED IN THE FOLLOWING WAY
C
C     DXDCR,DXDCI,XR,XI,DXDSR,DXDSI,DVDCR,DVDCI,VR,VI,DYDCR,DYDCI,YR,YI
C       1     2    3  4   5     6     7     8    9  10   11    12   13 14
C
      CALL RESETR(PC,14*16,0.0)
C
C   EQUILIBRIUM QUANTITIES
C
      Z1   =WNTORE*EQ(6,KCHI)
      ZNG  =WNTORE*EQ(11,KCHI)/QIAXE
      ZBETC=EQ(13,KCHI)
      ZH   =EQ(17,KCHI)
      ZDRS =EQ(15,KCHI)
      ZDRC =EQ(16,KCHI)
C
C     FIRST  SQUARE OF POTENTIAL ENERGY
C
      PC(1,1)=1.0
      PC(4,1)=0.0
      PC(2,2)=1.0
      PC(3,2)=0.0
C
C     SECOND SQUARE OF POTENTIAL ENERGY
C
      PC(4,3)=-Z1
      PC(5,3)=1.0
      PC(7,3)=1.0
      PC(10,3)=-ZNG
      PC(3,4)=Z1
      PC(6,4)=1.0
      PC(8,4)=1.0
      PC(9,4)=ZNG
C
C     THIRD  SQUARE OF POTENTIAL ENERGY
C
      PC(1,5)=ZBETC
      PC(3,5)=ZH
      PC(4,5)=-Z1
      PC(5,5)=1.0
      PC(10,5)=-ZNG
      PC(2,6)=ZBETC
      PC(3,6)=Z1
      PC(4,6)=ZH
      PC(6,6)=1.0
      PC(9,6)=ZNG
C
C     FOURTH SQUARE OF POTENTIAL ENERGY
C
      PC(3,7)=ZDRS
      PC(4,7)=-Z1
      PC(5,7)=1.0
      PC(7,7)=1.0
      PC(9,7)=ZDRC
      PC(10,7)=-ZNG
      PC(11,7)=1.0
      PC(13,7)=ZDRC
      PC(14,7)=0.0
      PC(3,8)=Z1
      PC(4,8)=ZDRS
      PC(6,8)=1.0
      PC(8,8)=1.0
      PC(9,8)=ZNG
      PC(10,8)=ZDRC
      PC(12,8)=1.0
      PC(13,8)=0.0
      PC(14,8)=ZDRC
C
C     FIFTH  SQUARE OF POTENTIAL ENERGY
C
      PC(3,9)=1.0
      PC(4,10)=1.0
C
C     FIRST SQUARE OF KINETIC ENERGY
C
      PC(3,11)=1.0
      PC(4,12)=1.0
C
C     SECOND SQUARE OF KINETIC ENERGY
C
      PC(3,13)=-ZBETC
      PC(9,13)=1.0
      PC(13,13)=1.0
      PC(4,14)=-ZBETC
      PC(10,14)=1.0
      PC(14,14)=1.0
C
C     THIRD SQUARE OF KINETIC ENERGY
C
      PC(13,15)=1.0
      PC(14,16)=1.0
      RETURN
      END
```

```
*DECK P3C2S10
      SUBROUTINE VECT2(PC,PF,PV)
C
C 3.2.10  COEFFICIENTS MULTIPLYING THE UNKNOWNS
C
      DIMENSION PC(14,16),PF(16),PV(16,16)
C
C     X-COMPONENT
C
      DO 100 J=1,16
      PV( 1,J)=PC( 1,J)*PF( 1)+PC( 3,J)*PF( 2)+PC( 5,J)*PF( 3)
      PV( 2,J)=PC( 2,J)*PF( 1)+PC( 4,J)*PF( 2)+PC( 6,J)*PF( 3)
      PV( 3,J)=PC( 1,J)*PF( 4)+PC( 3,J)*PF( 5)+PC( 5,J)*PF( 6)
      PV( 4,J)=PC( 2,J)*PF( 4)+PC( 4,J)*PF( 5)+PC( 6,J)*PF( 6)
      PV( 5,J)=PC( 1,J)*PF(11)+PC( 3,J)*PF(12)+PC( 5,J)*PF(13)
      PV( 6,J)=PC( 2,J)*PF(11)+PC( 4,J)*PF(12)+PC( 6,J)*PF(13)
      PV( 7,J)=PC( 1,J)*PF(14)+PC( 3,J)*PF(15)+PC( 5,J)*PF(16)
      PV( 8,J)=PC( 2,J)*PF(14)+PC( 4,J)*PF(15)+PC( 6,J)*PF(16)
C
C     Y-COMPONENT
C
      PV( 9,J)=PC(11,J)*PF( 7)+PC(13,J)*PF( 8)
      PV(10,J)=PC(12,J)*PF( 7)+PC(14,J)*PF( 8)
      PV(11,J)=PC(11,J)*PF( 9)+PC(13,J)*PF(10)
      PV(12,J)=PC(12,J)*PF( 9)+PC(14,J)*PF(10)
C
C     V-COMPONENT
C
      PV(13,J)=PC( 7,J)*PF( 7)+PC( 9,J)*PF( 8)
      PV(14,J)=PC( 8,J)*PF( 7)+PC(10,J)*PF( 8)
      PV(15,J)=PC( 7,J)*PF( 9)+PC( 9,J)*PF(10)
      PV(16,J)=PC( 8,J)*PF( 9)+PC(10,J)*PF(10)
 100  CONTINUE
C
      RETURN
      END
*DECK P3C2S11
      SUBROUTINE DIADC2(KV,PV,PC1)
C
C 3.2.11  DIADIC MULTIPLICATION
C
*CALL COMAUX
C
      DIMENSION PC1(16),PV(16,16)
C
      DO  200  K=1,10
      DO  100  I=1,KV
      DO  100  J=1,KV
      CONA(I,J)=CONA(I,J)+PC1(K)*PV(I,K)*PV(J,K)
 100  CONTINUE
 200  CONTINUE
      DO  400  K=11,16
      DO  300  I=1,KV
      DO  300  J=1,KV
      CONB(I,J)=CONB(I,J)+PC1(K)*PV(I,K)*PV(J,K)
 300  CONTINUE
 400  CONTINUE
C
      RETURN
      END
```

```
*DECK P3C2S12
      SUBROUTINE SYMTRY(K,NCHI,KCHI,KPSI)
C
C 3.2.12  IMPOSE SYMMETRY CONDITIONS
C
*CALL COMCON
*CALL COMPHY
C
      IF (KCHI.NE.1 ) GO TO 100
      ZSIGN=-1.0
      CALL TRANSF(K,16,1,3,+ZSIGN)
      CALL TRANSF(K,16,2,4,-ZSIGN)
      CALL TRANSF(K,16,5,7,+ZSIGN)
      CALL TRANSF(K,16,6,8,-ZSIGN)
      CALL TRANSF(K,16,9,11,-ZSIGN)
      CALL TRANSF(K,16,10,12,+ZSIGN)
      CALL TRANSF(K,16,13,15,-ZSIGN)
      CALL TRANSF(K,16,14,16,+ZSIGN)
  100 CONTINUE
      IF ( KCHI.NE.NCHI) GO TO 300
      IF ( .NOT.NLEINQ ) GO TO 200
C
C   SYMMETRY CONDITIONS AT PI FOR TRANSFORMED VARIABLES
C
      ZA=4.0*ASIN(1.0)*WNTORE*QTILDA(KPSI)
      ZC=COS(ZA)
      ZS=SIN(ZA)
      ZA1=4.0*ASIN(1.0)*WNTORE*QS(KPSI)
      ZA2=4.0*ASIN(1.0)*WNTORE*QS(KPSI+1)
      ZC1=COS(ZA1)
      ZC2=COS(ZA2)
      ZS1=SIN(ZA1)
      ZS2=SIN(ZA2)
      CALL TRAMAT(K,16,ZC,ZS,ZC1,ZS1,ZC2,ZS2)
      GO TO 300
  200 CONTINUE
C
C   SYMMETRY CONDITIONS AT PI FOR ORIGINAL UNKNOWNS
C
      CALL TRANSF(K,16,3 ,1 ,+ZSIGN)
      CALL TRANSF(K,16,4 ,2 ,-ZSIGN)
      CALL TRANSF(K,16,7 ,5 ,+ZSIGN)
      CALL TRANSF(K,16,8 ,6 ,-ZSIGN)
      CALL TRANSF(K,16,11, 9,-ZSIGN)
      CALL TRANSF(K,16,12,10,+ZSIGN)
      CALL TRANSF(K,16,15,13,-ZSIGN)
      CALL TRANSF(K,16,16,14,+ZSIGN)
  300 CONTINUE
C
      RETURN
      END
*DECK P3C2S13
      SUBROUTINE REGULA(K)
C
C 3.2.13    IMPOSE REGULARITY CONDITIONS
C
      CALL AWAY(K,16,1)
      CALL AWAY(K,16,2)
      CALL AWAY(K,16,3)
      CALL AWAY(K,16,4)
C
      RETURN
      END
*DECK P3C2S14
      SUBROUTINE BOUNDA(K)
C
C 3.2.14  IMPOSE BOUDARY CONDITIONS
C
      CALL AWAY(K,16,5)
      CALL AWAY(K,16,6)
      CALL AWAY(K,16,7)
      CALL AWAY(K,16,8)
C
      RETURN
      END
```

```
*DECK P3C2S15
      SUBROUTINE TRANSF(KMAT,L,I,J,S)
C
C 3.2.15  MATRIX TRANSFORMATION
C
*CALL COMAUX
C
C   MATRIX A
C
      IF (KMAT .EQ. 2) GO TO 200
C
C   ADD COLUMN I TO J
C
      DO 100 K=1,L
      CONA(J,K)=CONA(J,K)+S*CONA(I,K)
  100 CONTINUE
C
C   ADD ROW I TO J
C
      DO 110 K=1,L
      CONA(K,J)=CONA(K,J)+S*CONA(K,I)
  110 CONTINUE
      CALL AWAY(KMAT,L,I)
C
      RETURN
C
C   MATRIX B
C
  200 CONTINUE
C
C   ADD COLUMN I TO J
C
      DO 210 K=1,L
      CONB(J,K)=CONB(J,K)+S*CONB(I,K)
  210 CONTINUE
C
C   ADD ROW I TO J
C
      DO 220 K=1,L
      CONB(K,J)=CONB(K,J)+S*CONB(K,I)
  220 CONTINUE
      CALL AWAY(KMAT,L,I)
C
      RETURN
      END
*DECK P3C2S416
      SUBROUTINE TRAMAT(K,L,PC,PS,PC1,PS1,PC2,PS2)
C
C 3.2.16   PERFORM SYMETRY CONDITIONS FOR XT=X*EXP(I*N*G*CHI)
C
*CALL COMAUX
      DIMENSION U(16,16),UT(16,16)
C
C     FORM TRANSFORMATION MATRICES U,UT
C
      CALL RESETR(UT,256,0.0)
      DO  100  I=1,L
      UT(I,I)=1.0
 100  CONTINUE
      UT( 1, 3)=-PC1
      UT( 2, 3)= PS1
      UT( 1, 4)= PS1
      UT( 2, 4)= PC1
      UT( 5, 7)=-PC2
      UT( 6, 7)= PS2
      UT( 5, 8)= PS2
      UT( 6, 8)= PC2
      UT( 9,11)= PC
      UT(10,11)=-PS
      UT( 9,12)=-PS
      UT(10,12)=-PC
      UT(13,15)= PC
      UT(14,15)=-PS
      UT(13,16)=-PS
      UT(14,16)=-PC
      DO  110  I=1,L
      DO  110  J=1,L
      U(I,J)=UT(J,I)
 110  CONTINUE
      IF (K .EQ. 1) CALL MULT(UT,CONA,L,L)
      IF (K .EQ. 2) CALL MULT(UT,CONB,L,L)
      CALL MULT(UT,U,L,L)
      DO  200  I=1,L
      DO  200  J=1,L
      IF (K .EQ. 1) CONA(I,J)=UT(I,J)
      IF (K .EQ. 2) CONB(I,J)=UT(I,J)
 200  CONTINUE
      CALL AWAY(K,16,3)
      CALL AWAY(K,16,4)
      CALL AWAY(K,16,7)
      CALL AWAY(K,16,8)
      CALL AWAY(K,16,11)
      CALL AWAY(K,16,12)
      CALL AWAY(K,16,15)
      CALL AWAY(K,16,16)
      RETURN
      END
```

```
*DECK P3C2S17
      SUBROUTINE AWAY(K,L,KR)
C
C 3.2.17  REMOVE A COLOMN AND ROW
C
*CALL COMAUX
C---------------------------------------------------------------------
CL              1.              ELIMINATE A ROW/COLUMN OF A
C
      IF (K .EQ. 2) GO TO 200
      DO  110  I=1,L
      CONA(KR,I)=0.0
      CONA(I,KR)=0.0
 110  CONTINUE
      CONA(KR,KR)=1.0
C
      RETURN
C
C---------------------------------------------------------------------
CL              2.              ELIMINATE A ROW/COLUMN OF B
C
 200  CONTINUE
      DO  210  I=1,L
      CONB(KR,I)=0.0
      CONB(I,KR)=0.0
 210  CONTINUE
      CONB(KR,KR)=1.0E-20
C
      RETURN
      END
*DECK P3C2S18
      SUBROUTINE STORE(PA,PB)
C
C 3.2.18  ADD CELL CONTRIBUTION TO A OR B
C
*CALL COMAUX
*CALL COMNUM
*CALL COMESH
      DIMENSION PA(16,16),PB(16,16)
C
      NCON=16
      DO 100 I=1,16
 100  NPLAC(I)=NPLAC(I)+6*NCHI+6
      DO  150  JROW=1,16
      IROW=NPLAC(JROW)
      IF (IROW .EQ. 6*NCHI+6) GO TO 150
      DO  140  JCOL=1,16
      ICOL=NPLAC(JCOL)
      IF (ICOL .EQ. 6*NCHI+6) GO TO 140
      IC=ICOL-IROW
      IF (IC .LT. 0) GO TO 140
      IF (IC-4 .GT. 0) GO TO 110
C
C     THE FIRST 4 DIAGONALS
C
      AA(IC+1,IROW)=AA(IC+1,IROW)+PA(JROW,JCOL)
      BB(IC+1,IROW)=BB(IC+1,IROW)+PB(JROW,JCOL)
      GO TO 140
 110  CONTINUE
      IC=IC-2*NCHI+1
      IF (IC-7 .GT. 0) GO TO 120
C
C      THE FIRST 7 OUTER DIAGONAL BAND
C
      AA(IC+5,IROW)=AA(IC+5,IROW)+PA(JROW,JCOL)
      BB(IC+5,IROW)=BB(IC+5,IROW)+PB(JROW,JCOL)
      GO TO 140
 120  CONTINUE
      IC=IC-2*NCHI-2
      IF (IC-7 .GT. 0) GO TO 130
C
C      THE SECOND 7 OUTER DIAGONAL BAND
C
      AA(IC+12,IROW)=AA(IC+12,IROW)+PA(JROW,JCOL)
      BB(IC+12,IROW)=BB(IC+12,IROW)+PB(JROW,JCOL)
      GO TO 140
 130  CONTINUE
      IC=IC-2*NCHI-2
      IF (IC-7 .GT. 0) STOP "OUT OF RANGE"
C
C      THE THIRD 7 OUTER DIAGONAL BAND
C
      IR=IROW+4*NCHI+4
      AA(IC+12,IR)=AA(IC+12,IR)+PA(JROW,JCOL)
      BB(IC+12,IR)=BB(IC+12,IR)+PB(JROW,JCOL)
 140  CONTINUE
 150  CONTINUE
      RETURN
      END
```

```
*DECK P3C3S01
        SUBROUTINE IODSK3(K,KPSI)
C
C 3.3.1   HANDLES DISK FILES
C
*CALL COMPHY
*CALL COMEQU
*CALL COMAUX
*CALL COMESH
*CALL COMCON
*CALL COMOUT
*CALL NEWRUN
C
        GO TO (100,200,300,400,600,700) K
C
100     CONTINUE
C
C-----------------------------------------------------------------------
CL              1.              READ NAMELIST
C
        NSAVE = 8
        REWIND NSAVE
        READ (NSAVE,NEWRUN)
        RETURN
C
200     CONTINUE
C
C-----------------------------------------------------------------------
CL              2.              WRITE NAMELIST
C
        REWIND NSAVE
        WRITE(NSAVE,NEWRUN)
        RETURN
C
300     CONTINUE
C
C-----------------------------------------------------------------------
CL              3.              READ AND WRITE VACUUM CONTRIBUTION
C
        ICHI2=2*NCHI+2
        IEND=(ICHI2*(ICHI2+1))/2
        REWIND NVAC
        READ (NVAC)  (VAC(I),I=1,IEND)
        WRITE (NDA)  (VAC(I),I=1,IEND)
        RETURN
C
400     CONTINUE
C
C-----------------------------------------------------------------------
CL              4.              WRITE MATRIX BLOCKS OF A
C
        I6=6*NCHI+6
        WRITE (NDA)((AA(I,J),I=1,18),J=1,I6)
        WRITE (NDB)((BB(I,J),I=1,18),J=1,I6)
        IF (KPSI .NE. NPSI+1) RETURN
        I6=I6+1
        I8=8*NCHI+8
        WRITE (NDA)((AA(I,J),I=1,4),J=I6,I8)
        WRITE (NDB)((BB(I,J),I=1,4),J=I6,I8)
        IF (REXT .GT. 1.0 .AND. NLGREN) GO TO 300
        RETURN
C
600     CONTINUE
C
C-----------------------------------------------------------------------
CL              6.              REWIND DISK FILES
C
        REWIND NSCRTC
        REWIND NDA
        REWIND NDB
        RETURN
C
700     CONTINUE
C
C-----------------------------------------------------------------------
CL              7.              READ EQUILIBRIUM
C
        READ (MEQ)((EQ(I,J),I=1,20),J=1,NCHI)
        RETURN
        END
```

```
10.26.04.UCLP, IN, CLST01E,        1.221KLNS.          ** END OF LISTING **
```

References

Agmon, S. (1965): *Lectures on Elliptic Boundary Value Problems* (Van Nostrand Mathematical Studies No. 2, New York)

Appert, K., Berger, D., Gruber, R., Troyon, F., Rappaz, J. (1974a): Studium der Eigenschwingungen eines zylindrischen Plasmas mit der Methode der finiten Elemente, J. Appl. Math. Phys. **25**, 229–240

Appert, K., Gruber, R., Vaclavik, I. (1974b): Continuous spectra of a cylindrical magnetohydrodynamic equilibrium, Phys. Fluids **17**, 1471–1472

Appert, K., Berger, D., Gruber, R., Rappaz, J. (1975a): A new finite element approach to the normal mode analysis in magnetohydrodynamics, J. Comp. Phys. **18**, 284–299

Appert, K., Berger, D., Gruber, R., Troyon, F., Roberts, K.V. (1975b): THALIA – A one-dimensional MHD stability program using the method of finite elements, Comput. Phys. Commun. **10**, 11–29

Appert, K., Balet, B., Gruber, R., Troyon, F., Tsunematsu, T., Vaclavik, J. (1982): MHD computations for Alfvén wave heating in tokamaks, Nucl. Fusion **22**, 903–919

Artsimowitsch, L.A., Sagdejew, R.S. (1983): *Plasmaphysik für Physiker* (Teubner, Stuttgart)

Babuska, I., Aziz, A.K. (1972): "The Mathematical Foundations of the Finite Element Method with Applications to Partial Differential Equations", in *Survey Lectures on the Mathematical Foundations of the Finite Element Method*, ed. by A.K. Aziz (Academic, New York)

Barston, E.M. (1964): Electrostatic oscillations in inhomogeneous cold plasmas, Ann. Phys. **29**, 282–303

Barsukov, A.G., Kovrov, P.E., Kulygin, V.M. et al. (1983): "Investigation of Plasma Confinement and Injection Heating in the T-11 Tokamak", in *Proc. of the 9th Conf. on Plasma Phys. and Contr. Nucl. Fusion Research* (*Baltimore*, 1982), Vol. I (IAEA, Vienna) pp. 83–94

Bateman, G. (1978): *MHD Instabilities* (MIT Press, Cambridge, MA)

Bateman, G., Schneider, W., Grossmann, W. (1974): MHD instabilities as an initial boundary-value problem, Nucl. Fusion **14**, 669–683

Bateman, G., Peng, Y.-K.M. (1977): Magnetohydrodynamic stability of flux-conserving tokamak equilibria, Phys. Rev. Lett. **38**, 829–832

Bauer, F., Betancourt, O., Garabedian, P.R. (1978): *A Computational Method in Plasma Physics* (Springer Verlag, Berlin, Heidelberg, New York)

Berger, D. (1977): "Numerical Computations of the Ideal MHD Stability of Small Aspect Ratio Tokamaks", Thèse EPFL Lausanne, Switzerland

Berger, D., Gruber, R., Troyon, F. (1976): A finite element approach to the computation of the MHD spectrum of straight noncircular plasma equilibria, Comput. Phys. Commun. **11**, 313–323

Bernard, L.C., Dobrott, D., Helton, F.J., Moore, R.W. (1980): Stabilization of ideal MHD modes, Nucl. Fusion **20**, 1199–1206

Bernard, L.C., Helton, F.J., Moore, R.W. (1981): GATO: an MHD stability code for axisymmetric plasmas with internal separatrices, Comput. Phys. Commun. **24**, 377–380

Bernard, L.C., Helton, F.J., Moore, R.W., Todd, T.N. (1983): MHD beta limits: scaling laws and comparison with Doublet III data, Nucl. Fusion **23**, 1475–1484

Bernstein, I.B., Frieman, E.A., Kruskal, M.D., Kulsrud, R.M. (1958): An energy principle for hydromagnetic stability problems, Proc. Roy. Soc. A**244**, 17–40

Betancourt, O., Hernegger, F., Merkel, P., Nührenberg, J., Gruber, R., Troyon, F. (1983): Comparison of MHD stability results obtained with BETA 3D and HERA 2D codes, J. Comput. Phys. **52**, 187–197

Bineau, M. (1962): Stabilité hydromagnétique d'un plasma toroidal: Etude variationnelle de l'integrale d'énergie, Nucl. Fusion **2**, 130–147

Blum, J., Le Foll, J., Thooris, B. (1981): The self-consistent equilibrium and diffusion code SCED, Comput. Phys. Commun. **24**, 235–254

Boozer, A.H., Chu, T.K., Dewar, R.L., Furth, H.P., Goree, J.A., Johnson, J.L., Kulsrud, R.M., Monticello, D.A., Kuo-Petravic, G., Sheffield, G.V., Yoshikawa, S., Betancourt, O. (1982): "Two High-β Toroidal Configurations: A Stellerator and a Tokamak – Torsatron Hybrid", in *Proc. of 9th Int. Conf. on Plasma Physics and Contr. Nucl. Fusion Research* (*Baltimore*, 1982), Vol. 3 (IAEA, Vienna) pp. 129–139

Brackbill, J.U. (1976): Numerical magnetohydrodynamics for high-beta plasmas, in Meth. Comput. Phys. **16**, 1–41

Bramble, J., Osborn, J. (1973): Rate of convergence estimates for non-selfadjoint eigenvalue approximations, Math. Comp. **27**, 525–549

Burrell, K.H., Stambaugh, R.D., Angel, T.R. et al. (1983): Attainment of reactor level volume – averaged toroidal beta in Doublet III, Nucl. Fusion **23**, 536–540

Bussac, M.N., Pellat, R., Edery, D., Soulé, J.L. (1975): Internal kink modes in toroidal plasmas with circular cross sections, Phys. Rev. Lett. **35**, 1638–1641

Chance, M.S., Greene, J.M., Grimm, R.C., Johnson, J.L., Manickam, J., Kerner, W., Berger, D., Bernard, L.C., Gruber, R., Troyon, F. (1978): Comparative numerical studies of ideal magnetohydrodynamic instabilities, J. Comp. Phys. **28**, 1–13

Chandrasekhar, S. (1961): *Hydrodynamic and Hydromagnetic Stability*, Int. Series of Monographs on Physics (Clarendon, Oxford)

Charlton, L.A., Dory, R.A., Peng, Y.-K.M., Strickler, D.J., Lynch, S.J., Lee, D.K., Gruber, R., Troyon, F. (1979): Stability study of high-β flux-conserving equilibria, Phys. Rev. Lett. **43**, 1395–1398

Charlton, L.A., Lee, D.K. (1983): Study of the ideal MHD stability of helical equilibria, Plasma Phys. **25**, 1257–1269

Charlton, L.A., Lee, D.K., Wieland, R.M., Carreras, B.A., Cooper, W.A., Nielson, G.H. (1984): Ideal MHD stability of the ISX-B tokamak, Nucl. Fusion **24**, 33–38

Chen, F.F. (1974): *Introduction to Plasma Physics* (Plenum, New York)

Chodura, R., Schlüter, A. (1981): A 3D code for MHD equilibrium and stability, J. Comput. Phys. **41**, 68–88

Christiansen, J.P., Roberts, K.V. (1974): OLYMPUS, a standard control and utility package for initial-value FORTRAN programs, Comput. Phys. Commun. **7**, 245–270

Ciarlet, P.G. (1978): *The Finite Element Method for Elliptic Problems* (North-Holland, Amsterdam)

Connor, J.W., Hastie, R.J., Taylor, J.B. (1978): Shear, periodicity, and plasma ballooning modes, Phys. Rev. Lett. **40**, 396–399

Correa-Restrepo, D. (1978): Ballooning modes in three-dimensional MHD equilibria, Z. Naturforsch. A **33**, 789–791

Correa-Restrepo, D. (1982): Resistive ballooning modes in three-dimensional configurations, Z. Naturforsch. A **37**, 848–858

Degtyarev, L.M., Drozdov, V.V., Martynov, A.A., Medvedev, S.Yu. (1984): "Ideal MHD Beta Limits in Tokamak", in *Proc. of Invited Papers of the International Conf. on Plasma Physics (Lausanne)*, Vol. I, p. 157

Descloux, J. (1979): Error bounds for an isolated eigenvalue obtained by the Galerkin method, J. Appl. Math. Phys. **30**, 167–176

Descloux, J. (1981): Essential numerical range of an operator with respect to a coercive form and the approximation of its spectrum by the Galerkin method, SIAM J. Numer. Anal. **18**, 1128–1133

Descloux, J., Nassif, N., Rappaz, J. (1977): "Spectral Approximations with Error Bounds for Non-Compact Operators', Rapport Dept. de Mathématiques EPFL

Descloux, J., Nassif, N., Rappaz, J. (1978a): On spectral approximation. Part 1. The problem of convergence, RAIRO Analyse Numérique **12**, 97–112

Descloux, J., Nassif, N., Rappaz, J. (1978b): On spectral approximation. Part 2. Error estimates for the Galerkin method, RAIRO Analyse Numérique **12**, 113–119

Descloux, J., Nassif, N., Rappaz, J. (1979): *On Properties of Spectral Approximation*, Lecture Notes Math., Vol. 703 (Springer, Berlin, Heidelberg, New York) pp. 81–85

Descloux, J., Luskin, M., Rappaz, J. (1981): Approximation of the spectrum of closed operators: the determination of normal modes of a rotating basin, Math. Comp. **36**, No. 153, 137–154

Dewar, R.L., Grimm, R.C., Johnson, J.L., Frieman, E.A., Greene, J.M., Rutherford, P.H. (1974): Long-wavelength kink instabilities in low-pressure, uniform axial current, cylindrical plasmas with elliptic cross sections, Phys. Fluids **17**, 930–938

D'Ippolito, D.A., Freidberg, J.P., Goedbloed, J.P., Rem, J. (1978): High-beta tokamaks surrounded by force-free fields, Phys. Fluids **21**, 1600–1616

Dobrott, D., Nelson, D.B., Greene, J.M., Glasser, A.H., Chance, M.S., Frieman, E.A. (1977): Theory of ballooning modes in tokamaks with finite shear, Phys. Rev. Lett. **39**, 943–946

Evequoz, H. (1980): "Approximation spectrale liée à l'étude de la stabilité MHD d'un plasma par une méthode d'élèments finis non conformes", Thése No. 375, EPF-Lausanne

Evequoz, H., Jaccard, J. (1980): A nonconforming finite element method to compute the spectrum of an operator relative to the stability of a plasma in toroidal geometry, Numer. Math. **36**, 455–465

Edery, D., Pellat, R., Soulé, J.L. (1981): Effects of toroidal coupling on the non-linear evolution of tearing modes and on the stochastisation of the magnetic field topology in plasmas, Comput. Phys. Commun. **24**, 427–436

Festeau-Barrioz, M.C., Weibel, E.S. (1982): Large amplitude solutions of the nonlinear wave equation for an ideal, cold three-fluid plasma, Comput. Phys. Commun. **27**, 11–23

Finan III, C.H., Killeen, J. (1981): Solution of the time dependent, three-dimensional resistive magnetohydrodynamic equations, Comput. Phys. Commun. **24**, 441–463

Finn, J.M., Reimann, A. (1982): Tilt and shift mode stability in spheromaks with line tying, Phys. Fluids **25**, 116–125

Fix, G. (1973): Eigenvalue approximation by the finite element method, Adv. Math. **10**, 300–316

Furth, H.P., Killeen, J., Rosenbluth, M.N., Coppi, B. (1966): "Stabilization by Shear and Negative V", in *Proc. of Plasma Physics and Contr. Nucl. Fusion Research*, Vol. 1 (IAEA, Vienna) pp. 103–126

Gautier, P., Gruber, R., Troyon, F. (1981): Numerical study of the ideal-MHD stability limits of oblate spheromaks, Nucl. Fusion **21**, 1399–1407

Goedbloed, J.P. (1975): Spectrum of ideal magnetohydrodynamics of axisymmetric toroidal systems, Phys. Fluids **18**, 1258–1268

Goedbloed, J.P. (1981): Conformal mapping methods in two-dimensional magnetohydrodynamics, Comput. Phys. Commun. **24**, 311–321

Goedbloed, J.P., Hagebeuk, H.J.L. (1972): Growth rates of instabilities of a diffuse linear pinch, Phys. Fluids **15**, 1090–1101

Goedbloed, J.P., Sakanaka, P.H. (1974): New approach to magnetohydrodynamic stability, Phys. Fluids **17**, 908–929

Greene, J.M., Johnson, J.L. (1962): Stability criterion for arbitrary hydromagnetic equilibria, Phys. Fluids **5**, 510–517

Grimm, R.C., Greene, J.M., Johnson, J.L. (1976): Computation of the MHD spectrum in axisymmetric toroidal confinement systems, Meth. Comput. Phys. **16**, 253–280

Grubb, G., Geymonat, G. (1977): The essential spectrum of elliptic systems of mixed order, Math. Ann. **227**, 247–276

Gruber, R. (1978): Finite hybrid elements to compute the ideal MHD spectrum of an axisymmetric plasma, J. Comp. Phys. **26**, 379–389

Gruber, R. (1980): Hymniablock: eigenvalue solver for blocked matrices, Comput. Phys. Commun. **20**, 421–428

Gruber, R., Troyon, F. (1977): "Calculs de stabilité MHD d'un plasma par la méthode variationnelle", in *Computing Methods in Applied Sciences and Engineering*, Lecture Notes Phys., Vol. 91, ed. by R. Glowinski, J.L. Lions (Springer, Berlin, Heidelberg, New York) pp. 288–301

Gruber, R., Troyon, F., Berger, D., Bernard, L.C., Rousset, S., Schreiber, R., Kerner, W., Schneider, W., Roberts, K.V. (1981a): ERATO stability code, Comput. Phys. Commun. **21**, 323–371

Gruber, R., Semenzato, S., Troyon, F., Tsunematsu, T., Kerner, W., Merkel, P., Schneider, W. (1981b): HERA and other extensions of ERATO, Comput. Phys. Commun. **24**, 363–376

Gruber, R., Pfersisch, Ch., Semenzato, S., Troyon, F., Tsunematsu, T. (1981c): On the numerical determination of ideal MHD limits of stability of axisymmetric toroidal configurations, Comput. Phys. Commun. **24**, 381–387

Gruber, R., Kerner, W., Merkel, P., Nührenberg, J., Schneider, W., Troyon, F. (1981d): Stability-beta limit of helical equilibria, Comput. Phys. Commun. **24**, 389–398

Hain, K., Lüst, R., Schlüter, A. (1957): Z. Naturforsch. A **12**, 833

Hain, K., Lüst, R. (1958): Zur Stabilität zylindersymmetrischer Plasmakonfigurationen mit Volumenströmen, Z. Naturforsch. A **13**, 936–940

Hender, T.C., Robinson, D.C. (1981): Linear and nonlinear resistive instability studies, Comput. Phys. Commun. **24**, 413–419

Jaccard, J. (1980): "Approximation spectrale par la méthode des élèments finis conformes d'une classe d'opèrateurs non compacts et partièllement réguliers", Thése EPF-Lausanne, No. 374

Jaccard, J., Evequoz, H. (1982): Spectral approximation of the spectrum of an operator given by the MHD-stability of a plasma", Math. Comput. **39**, No. 160, 443–452

Jardin, S.C. (1982): "Ideal magnetohydrodynamic stability of the spheromak configuration", Nucl. Fusion **22**, 629–642

Johnson, D., Bell, M., Bitter, M. et al. (1983): "High-Beta Experiments with Neutral-Beam Injection on PDX", in *Proc. of the 9th Conf. on Plasma Phys. and Contr. Nucl. Fusion Research* (*Baltimore*, 1982), Vol. I (IAEA, Vienna) pp. 9–26

Kadomtsev, B.B. (1966): "Hydromagnetic Stability of a Plasma", in *Reviews of Plasma Physics*, Vol. 2, ed. by M.A. Leontovich (Consultants Bureau, New York) pp. 153–199

Kerner, W. (1976): Numerical study of the MHD spectrum in tokamaks with a non-circular cross-section, Nucl. Fusion **16**, 643–650

Kerner, W., Tasso, H. (1975): "Kink Instabilities for Shaped Tokamaks in Toroidal Geometry", in *Proc. of Plasma Physics and Contr. Nucl. Fusion Research* (*Tokyo*, 1974), Vol. I, pp. 475–484

Kerner, W., Gruber, R., Troyon, F. (1980): Numerical study of the internal kink mode in tokamaks, Phys. Rev. Lett. **44**, 536–540

Kerner, W., Gautier, P., Lackner, K., Schneider, W., Gruber, R., Troyon, F. (1981): Ideal magnetohydrodynamic stability of high-beta, high-current tokamak equilibria, Nucl. Fusion **21**, 1383–1397

Kerner, W., Lerbinger, K., Gruber, R., Tsunematsu, T. (1984): Normal mode analysis for linear resistive Magnetohydrodynamics, IPP Report 6, 235 and to be published in Computer Phys. Commun.

Laval, G., Pellat, R., Soulé, J.L. (1974): Hydromagnetic stability of a current-carrying pinch with noncircular cross section, Phys. Fluids **17**, 835–845

Lefèvre, D., Rappaz, J. (1978): "Sur la stabilité d'un écoulement de contre-courant", Ecole Polytéchnique de Paris, Centre de Mathématique Appliqué, Rapport interne No. 37

Leonov, V.M., Merezhkin, V.G., Mukhovatov, V.S., Sannikov, V.V., Tiliuin, G.N. (1981): "Ohmic-Heating and Neutral-Beam Injection Studies on the T-11 Tokamak", in *Proc. of the 8th Conf. on Plasma Phys. and Contr. Nucl. Fusion Research* (*Brussels*, 1980), Vol. I (IAEA, Vienna) pp. 393–403

Lundquist, S. (1951): On the stability of magneto-hydrostatic fields, Phys. Rev. **83**, 307

Lüst, R., Schlüter, A. (1954): Z. Astrophys. **34**, 263

Lynch, V.E., Carreras, B.A., Hicks, H.R., Holmes, J.A., Garcia, L. (1981): Resistive MHD studies of high β tokamak plasmas, Comput. Phys. Commun. **24**, 465–476

Lyon, J.F., Carreras, B.A., Harris, J.H. et al. (1983): "Stellarator Physics Evaluation Studies", in *Proc. of the 9th Conf. on Plasma Phys. and Contr. Nucl. Fusion Research* (*Baltimore*, 1982), Vol. III (IAEA, Vienna) pp. 115–126

Manickam, J. (1984): Stability of $n=1$ internal modes in tokamaks, Nucl. Fusion **24**, 595–607

Manickam, J., Grimm, R.C., Dewar, R.L. (1981): The linear stability analysis of MHD models in axisymmetric toroidal geometry, Comput. Phys. Commun. **24**, 355–361

Mercier, C. (1962): Un critère de stabilité d'un système toroidal hydromagnétique en pression scalaire, Nucl. Fusion Suppl. **2**, 801–808

Merkel, P. (1982): A Green's function method for the vacuum contribution to the MHD stability of helically symmetric equilibria, Z. Naturforsch. A **37**, 859–866

Merkel, P., Nührenberg, J., Gruber, R., Troyon, F. (1983): Linear MHD stability studies of helically symmetric equilibria with HERA, Nucl. Fusion **23**, 1061–1069

Merkel, P., Nührenberg, J. (1984): HASE – a quasi-analytical 2D MHD equilbrium code, Comput. Phys. Commun. **31**, 115–122

Mills, W. (1979): The resolvent stability condition for spectra convergence with application to the finite element approximation of noncompact operators, SIAM J. Numer. Anal. **16**, 695–703

Miyamoto, K. (1980): *Plasma Physics for Nuclear Fusion* (MIT Press, Cambridge, MA)

Morris, M., Todd, T.N. (1983): Private communications

Murakami, M., Swain, D.W., Bates, S.C. et al. (1981): "Neutral-Beam Injection Experiments in the ISX-B Tokamak", in *Plasma Physics and Controlled Nuclear Fusion Research*, Proc. 8th Int. Conf., Brussels, 1980, Vol. 1 (IAEA, Vienna) pp. 377–392

Nagami, M. and the JAERI Team, D. Overskei and the GA Team (1983): "High-β Injection Experiments with Shaped Plasmas in Doublet-III", in *Proc. of the 9th Conf. on Plasma Phys. and Contr. Nucl. Fusion Research* (*Baltimore*, 1982), Vol. I (IAEA, Vienna) pp. 27–40

Newcomb, W.A. (1960): Hydromagnetic stability of a diffuse linear pinch, Ann. Phys. **10**, 232–267

Neilson, G.H., Lazarus, E.A., Murakami, M. et al. (1983): Beta and confinement scaling studies with neutral beam heating in the ISX-B tokamak, Nucl. Fusion **23**, 285–294

Okabayashi, M., Todd, A.M.M. (1980): A numerical study of MHD equilibrium and stability of the spheromak, Nucl. Fusion **20**, 571–577

Osborn, J. (1975): Spectral approximation for compact operators, Math. Comp. **29**, 712–725

Pfersich, C., Gruber, R., Troyon, F. (1983): Free-boundary MHD stability of pressureless oblate spheromaks-dependence on aspect ratio and elongation, Nucl. Fusion **23**, 1127–1134

Platzman, G.W. (1975): Normal modes of the atlantic and indian oceans, J. Phys. Oceanogr. **5**, 201–221

Rappaz, J. (1976): "Approximation par la méthode des élèments finis du spectre d'un opérateur non-compact donné par la stabilité MHD d'un plasma", Thèse No. 239 (EPF-Lausanne)

Rappaz, J. (1977): Approximation of the spectrum of a noncompact operator given by the magnetohydrodynamic stability of a plasma, Numer. Math. **28**, 15–24

Rappaz, J. (1979): "Spectral Approximation by Finite Elements of a Problem of MHD Stability of a Plasma", in *The Mathematics of Finite Elements and Applications*, ed. by J.R. Whiteman (Academic, New York) pp. 311–318

Rappaz, J. (1982): "Some Properties on the Stability Related to the Approximation of Eigenvalue Problems", in *Computing Methods in Applied Sciences and Engineering V*, ed. by R. Glowinski, J.L. Lions (North-Holland, Amsterdam) pp. 167–175

Richtmyer, R.D., Morton, K.W. (1967): *Difference Methods for Initial-Value Problems*, 2nd ed. (Interscience, New York)

Roberts, K.V. (1974): An introduction to the OLYMPUS system, Comput. Phys. Commun. **7**, 237–243

Rosenbluth, M.N., Bussac, M.N. (1979): MHD stability of spheromak, Nucl. Fusion **19**, 489–498

Scott, D.S., Gruber, R. (1981): Implementing sparse matrix techniques in the ERATO code, Comput. Phys. Commun. **23**, 115–121

Schlüter, A., Schwenn, U. (1981): Equilibrium and stability studies with the 3D MHD code TUBE, Comput. Phys. Commun. **24**, 263–300

Schmidt, G. (1966): *Physics of High Temperature Plasmas* (Academic, New York)

Schwarz, H.R. (1980): *Methode der Finiten Elemente* (Teubner, Stuttgart)

Semenzato, S., Gruber, R. Zehrfeld, H.P. (1984): Computation of symmetric ideal MHD flow equilibria, Comput. Phys. Reports **1**, 389–425

Shafranov, V.D. (1970): Hydromagnetic Stability of a Current-Carrying Pinch in a Strong Longitudinal Magnetic Field, Soviet Physics – Technical Physics **15**, 175–183

Stix, T.H. (1962): *The Theory of Plasma Waves* (Mc Graw-Hill, New York)

Strang, G., Fix, G.J. (1973): *An Analysis of the Finite Element Method* (Prentice Hall, Englewood Cliffs, NJ)

Suydam, B.R. (1958): "Stability of a Linear Pinch", in *Proc. of the 2nd United Nations Int. Conf. on the Peaceful Uses of Atomic Energy* (*Geneva*, 1958), Vol. 31, pp. 157–159

Suzuki, N., Imai, T., Fujisawa, N. et al. (1981): "Recent Results on the Modified JFT-2 Tokamak", in *Proc. of the 8th Conf. on Plasma Phys. and Contr. Nucl. Fusion Research* (*Brussels*, 1980), Vol. II (IAEA, Vienna) pp. 525–533

Swain, D.W., Murakami, M., Bates, S.C. et al. (1981): High-beta injection experiments on the ISX-B tokamak, Nucl. Fusion **21**, 1409–1423

Sykes, A., Turner, M.F., Patel, S. (1983): "Beta Limits in Tokamaks Due to High-N Ballooning Modes", in *Proc. of 11th European Conf. on Controlled Fusion and Plasma Physics* (*Aachen*, 1983), Vol. II, pp. 363–366

Sykes, A., Wesson, J. (1974): Two-dimensional calculation of Tokamak stability, Nucl. Fusion **14**, 645–648

Sykes, A., Wesson, J.A. (1980): Major disruptions in tokamaks, Phys. Rev. Lett. **44**, 1215–1218

Takeda, T., Shimomura, Y., Ohta, M., Yoshikawa, M. (1972): Numerical analysis of MHD instabilities by the finite element method, Phys. Fluids **15**, 2193–2201

Takeda, T., Tsunematsu, T. (1979): "A numerical code SELENE to calculate axisymmetric toroidal MHD equilibria", JAERI-M 8042

Takeda, T., Tsunematsu, T., Tuda, T., Azumi, M., Takizuka, T., Tokuda, S., Kurita, G., Itoh, K., Naraoka, K., Tanaka, Y., Itoh, S.-I. (1983): "Physics of Intensively Heated Tokamak Plasmas", in *Proc. of 9th Conf. on Plasma Phys. and Controlled Nuclear Fusion Research* (*Baltimore*, 1982), Vol. III (IAEA, Vienna) p. 23–32

Takizuka, T., Tokuda, S., Azumi, M., Takeda, T. (1981a): Effects of the finite hybrid element on MHD stability calculations in a cylindrical plasma, Comput. Phys. Commun. **23**, 19–26

Takizuka, T., Tsunematsu, T., Tokuda, S. et al. (1981b): "Computational studies of tokamak plasmas", JAERI-M 9354

Todd, A.M.M., Manickam, J., Okabayashi, M., Chance, M.S., Grimm, R.C., Greene, J.M., Johnson, J.L. (1979): Dependence of ideal MHD kink and ballooning modes on plasma shape and profiles in tokamaks, Nucl. Fusion **19**, 743–752

Troyon, F., Gruber, R., Saurenmann, H., Semenzato, S., Succi, S. (1983): MHD limits to plasma confinement, Plasma Phys. **26**, 1A, 209–215

Troyon, F., Gruber, R. (1984): "A scaling law for the β-limit in tokamaks", Lausanne Report LRP 239/84 and submitted to Phys. Letters

Uo, K., Hyoshi, A., Obiki, T. et al. (1981): "Recent Developments in Heliotron Research", in *Proc. 8th Int. Conf. on Plasma Phys. and Controlled Nucl. Fusion Research* (*Brussels*, 1980), *Vol. I* (IAEA, Vienna) pp. 217–224

Wagner, F., Becker, G., Behringer, K. et al. (1983): "Confinement and β_p-studies in neutral injection heated Asdex plasmas", Max Planck Institute Report, IPP III 186

Wakatani, M., Yoshioka, T., Hanatani, K., Motojima, O., Iiyoshi, W., Uo, K. (1979): Magnetohydrodynamic instabilities in a high shear helical system, J. Phys. Soc. Jpn **47**, 974–983

Weibel, E.S. (1968): Dimensionally correct transformations between different systems of units, Am. J. Phys. **36**, 1130–1133

Wesson, J.A. (1975): "Hydromagnetic Stability of Tokamaks", in *Proc. of 7th European Conf. on Controlled Fusion and Plasma Physics, Lausanne*, Vol. II, pp. 102–118

Yamamoto, S., Maeno, M., Suzuki, N. et al. (1981): Magnetohydrodynamic activity in the JFT-2 tokamak with high-power neutral-beam-injection heating, Nucl. Fusion **21**, 993–1003

Zienkiewicz, D.C. (1977): *The Finite Element Method*, 3rd ed. (McGraw-Hill, London)

Subject Index